T0314298

# The Physical Nature of Information

# The Physical Nature of Information

A SHORT COURSE

## GREGORY FALKOVICH

ILLUSTRATIONS BY
NATALIA VLADIMIROVA

PRINCETON UNIVERSITY PRESS
PRINCETON AND OXFORD

Copyright © 2025 by Princeton University Press

Princeton University Press is committed to the protection of copyright and the intellectual property our authors entrust to us. Copyright promotes the progress and integrity of knowledge created by humans. Thank you for supporting free speech and the global exchange of ideas by purchasing an authorized edition of this book. If you wish to reproduce or distribute any part of it in any form, please obtain permission.

Requests for permission to reproduce material from this work should be sent to permissions@press.princeton.edu

Published by Princeton University Press
41 William Street, Princeton, New Jersey 08540
99 Banbury Road, Oxford OX2 6JX

press.princeton.edu

All Rights Reserved

ISBN 9780691266534
ISBN (e-book) 9780691266541

British Library Cataloging-in-Publication Data is available

Editorial: Abigail Johnson
Production Editorial: Sara Lerner
Jacket Design: Wanda España
Production: Danielle Amatucci
Publicity: William Pagdatoon
Copyeditor: Jennifer McClain
Jacket images: Rostislav Zatonskiy / Artur Golbert / Alamy Stock Photo

This book has been composed in Arno Pro

Printed in the United States of America

10 9 8 7 6 5 4 3 2 1

In memory of Leo Szilard and John Wheeler

# CONTENTS

# PREFACE

This book answers the eternal question, *How much can we say and do about something we do not know?* People in many trades and walks of life have perfected the art of bluffing without blushing, and acting blindly. This particular text grew out of a one-semester course, intended as a parting gift to those leaving physics for greener pastures and wondering what was worth taking with them. Statistically, most former physicists use statistics because this discipline was the first to develop quantitative tools to answer the question posed above. Yet when the course was taught in different institutions and countries, it attracted a motley mix of students, postdocs, and faculty from physics, mathematics, engineering, computer science, economics, and biology. Eventually, it evolved into a meeting place where we learn from each other, using the universal language of information theory.

The simplest way to answer the opening question is with a phenomenology traditionally called thermodynamics. It deals only with visible manifestations of the hidden, using general principles (like symmetries and conservation laws) to restrict possible outcomes. The focus is on mean values, and fluctuations are ignored. A more sophisticated approach derives the statistical laws by explicitly averaging over the hidden degrees of freedom. Those laws justify thermodynamics and describe the probability of fluctuations. Two basic notions of this approach—entropy and free energy—turn out to be among the few most important conceptual and technical tools of modern science and technology.

This book is an introduction that requires prior knowledge of neither thermodynamics and statistics nor information theory. The first chapter gives, in a minimalist way, the basics of thermodynamics and statistical physics and describes their dual focus on what we have (energy) and what we don't (knowledge). When ignorance exceeds knowledge, the right strategy is to measure ignorance. Entropy does that. We understand that *entropy is not a property of a system, but the information we lack about it.* It is then natural to start using in the second chapter the language of information theory, revealing the universality of the approach. When viewed from the perspective of information theory, the essence of statistics is essentially common sense, which could be compressed to the maxim "the whole truth and nothing but the truth." That means using all the available data and maximizing missing information, that is, looking for the entropy maximum conditional on the data. This approach is valid not only for thermal equilibrium (where data are on conserved quantities) but for any state. Mathematically, the approach is based to a large extent on the simple trick of adding many random numbers. Building on that basis, one develops versatile instruments, like mutual information and its quantum sibling, entanglement entropy, which are widely applied to subjects ranging from bacteria and neurons to markets and quantum computers. The third chapter describes several applications, elucidating different aspects of the approach and directions of its development. The fourth chapter follows a tireless random walker to rank web pages and obtain a more powerful form of the second law of thermodynamics. In the fifth chapter, we learn that only full knowledge must persist; if we let in even the smallest semblance of ignorance, it grows and fills the space. This is illustrated by the irreversible entropy change produced by reversible flows in phase space. We also discuss the most sophisticated way to forget information: the renormalization group. Forgetting is a fascinating activity—one learns truly important things this way. The sixth chapter describes the fundamental lower and upper limits of uncertainty imposed, respectively, by quantum theory and relativity. Chapter 7 closes out the main body of the text with a compact conclusion stating the main lessons. The appendix presents more advanced subjects and material for tutorials. Exercises are provided after the respective sections. The last section of the appendix lists all the exercises presented in the book, along with detailed solutions.

The textbook is self-sufficient and contains no reference list. Those wishing to go above and beyond are encouraged to do searches using prompts like (Joule 1845), which will bring you not only to the original text but also to important texts citing it.

Even though this is a graduate text, we use only elementary mathematical tools, but from all three fields—geometry, algebra, and analysis—which correspond, respectively, to studying space, time, and continuum. We employ two complementary ways of thinking: continuous flows and discrete counting (thus involving both brain hemispheres). They equip the reader with a powerful and universal tool, applied everywhere, from computer science and machine learning to biophysics and economics. The book is panoramic, trying to combine into a reasonably coherent whole the subjects taught in much detail in different departments: thermodynamics and statistical mechanics (as taught in physics and engineering), dynamical chaos (as taught in physics and applied mathematics), information and communication theories (as taught in computer science and engineering). My desire is to reveal an essential unity between these disciplines. In addition, I felt compelled to tell a story worth telling: how we discover the limits imposed by uncertainty on engines, communications, computations, and perception. The story's protagonist is the notion of entropy/information, which was born in the Industrial Revolution, matured during the digital revolution, and leads the present revolution, blurring the boundaries between physical, digital, and biological domains.

In the end, recognizing the informational nature of the world and breaking the barriers of specialization is also of value for physicists. People working on quantum computers and the entropy of black holes use the same tools as those designing self-driving cars and market strategies, studying molecular biology, animal behavior, and human languages, figuring out how the brain works, and trying to quantify conscience. Many go out and apply the tools of physics to new domains. Few can come back enriched by the knowledge of how the tools work in linguistics and brain research and look at the physical theories as an example of human language developed by the human brain. It may open new perspectives.

The amount of material exceeds that for a standard one-semester course so that lecturers can choose what is more appropriate for their audience. Shorter and less technical versions of the course can be based on chapters 1–4 and 7. A longer course can include chapters 5 and 6, which involve more physics. The book can also be used for independent study by senior undergraduate and graduate students, postdocs, and faculty who want to see a bigger picture with connections between different disciplines and find new research opportunities. Readers familiar with thermodynamics and statistics can start from the second chapter, consulting the material from the first one when it is referred to. On the other hand, readers from computer science, engineering, mathematics, or biology may benefit from reading the first chapter, as it provides a unifying framework for the rest of the book. Bear in mind that the book is written by a natural scientist focused more on "how it works" and "what it is like" and less on the rigor of definitions and statements. Truly impatient readers could read only the short seventh chapter, which lists the take-home lessons.

For a book with such a wide scope, it is probably inevitable not only that my limited expertise in engineering, computer science, biology, economics, and linguistics has caused infelicities and technical errors but that the dilettante perspective has distorted essential elements in the culture of these disciplines. As Schrodinger wrote, "Some of us should venture to embark on a synthesis of facts and theories, albeit with second-hand and incomplete knowledge of some of them—and at the risk of making fools of ourselves." Fully accepting this risk, I shall maintain a website at https://www.weizmann.ac.il/complex/falkovich/information-theory where objections and corrections will be gratefully received and discussed.

Small-print parts can be omitted upon the first few readings.

*Gregory Falkovich, 2024*
Rehovot, Israel

"Entropy is not a property of a system, but the information we lack about it, right? Where is my Information textbook?"

# ACKNOWLEDGMENTS

I wish to thank my friends and colleagues, whose expert advice was extremely helpful: Yakov Sinai, Alexander Zamolodchikov, Eric Siggia, Leonid Levitov, Massimo Vergassola, Alexei Kitaev, Jorge Kurchan, Boris Shraiman, William Bialek, Anton Kapustin, David Khmelnitskii, and Krzysztof Gawedzki. I am grateful to students, postdocs, and faculty who attended the lectures at the Weizmann Institute, Harvard, New York University, the Simons Center, University of Arizona, Skoltech, and the High School of Economics, and whose feedback was invaluable in writing this text. Special thanks to Michal Shavit at Weizmann and Saranesh Prembabu at Harvard for helping with tutorials, problems, and solutions and for catching numerous errors (the remaining ones are my responsibility alone). I am grateful to Natalia Vladimirova for the cover art, illustrations, and help with the figures.

# The Physical Nature of Information

# 1

# Thermodynamics and Statistical Physics

Our knowledge is always partial. If we study macroscopic systems, some degrees of freedom remain hidden. For small sets of atoms or subatomic particles, their quantum nature prevents us from knowing precise values of their momenta and coordinates simultaneously. We used to believe that we found the way around the partial knowledge in mechanics, electricity, and magnetism, where we have *closed sets of equations describing explicitly known degrees of freedom*. In other words, we learned how to restrict our description only to things that can be considered independent of the unknown within a given accuracy. For example, planets are large complex bodies, and yet the motion of their centers of mass in the limit of large distances satisfies closed equations.[1]

Despite the spectacular successes of electromagnetic theory and celestial mechanics, we soon realized how illusory was our belief in the closed description, since we needed to feed it with initial or boundary conditions taken from measurements. Here our knowledge is incomplete because of a finite precision of measurements. This has dramatic consequences when there is instability, so small data uncertainty at a given moment leads to large uncertainty in predicting the future and recovering the past. In a sense, every new decimal in precision is a new degree of freedom for unstable systems (including our solar system).

In this chapter, we deal with *observable manifestations of the hidden degrees of freedom*. While we do not know their state, we do know their nature—whether those degrees of freedom are related to moving particles, spins, bacteria, or

---

1. The next step—description of a planet rotation—needs to account for many extra degrees of freedom, for instance, oceanic flows (which slow down rotation by tidal forces).

market traders. That means, in particular, that we know the symmetries and conservation laws of the system.

The first two sections present a phenomenological approach called thermodynamics. The last two sections serve as a brief introduction into statistical physics.

## 1.1   Basics of Thermodynamics

One can teach monkeys to differentiate; integration requires humans.

—GLEB KOTKIN

For at least a few thousand years, people have been burning things to propel objects. That was put on an industrial scale by the use of steam engines in the mid- to late 1700s. The Industrial Revolution generated a practical need to estimate engine efficiency, which triggered a regular scientific inquiry on general principles governing the conversion of heat into mechanical work. That led to the development of the abstract concept of entropy.

A heat engine works by delivering heat from a reservoir with some temperature $T_1$ via some system to another reservoir, with $T_2$, doing some work in the process. Look under the hood of your car to appreciate the level of abstraction achieved in that definition. The work $W$ is the difference between the heat given by the hot reservoir, $Q_1$, and the heat absorbed by the cold one, $Q_2$. What is the maximal fraction of heat we can use for work? Carnot in 1824 stated that we cannot make $Q_2$ arbitrarily small: in all processes, $Q_2/T_2 \geq Q_1/T_1$, so that the efficiency is bounded from above:

$$\frac{W}{Q_1} = \frac{Q_1 - Q_2}{Q_1} \leq 1 - \frac{T_2}{T_1}. \tag{1.1}$$

His elaborate arguments are of only historical interest now. Clausius in 1864 called the ratio $Q/T$ *entropy* (the word starts with *en-*, like *energy*, and ends with *-tropos*, which means "turn" or "way" in Greek). We now interpret the Carnot criterion, saying that the entropy decrease of the hot reservoirs, $\Delta S_1 = Q_1/T_1$, must be less than the entropy increase of the cold one, $\Delta S_2 = Q_2/T_2$. Maximal work is achieved for minimal (zero) total entropy change, $\Delta S_2 = \Delta S_1$.

Just like the path from the Carnot engine to general thermodynamics, we discover the laws of nature by induction: from particular cases to general law and from processes to state functions. The latter step requires integration (to pass, for instance, from the Newton mechanics equations to the Hamiltonian or from thermodynamic equations of state to thermodynamic potentials). It is much easier to differentiate than to integrate, so deduction (or the postulation approach) is usually more pedagogical.[2] It also provides a good vantage point for generalizations and appeals to our brain, which likes to hypothesize before receiving any data, as we shall see later. In such an approach, one starts by postulating a variational principle for some function of the state of the system. Then one deduces from that principle the laws that govern change when one passes from state to state.

Here we present a deductive description of thermodynamics. *Thermodynamics studies restrictions on the possible macroscopic properties that follow from the fundamental conservation laws.* Therefore, thermodynamics does not predict numerical values but rather sets inequalities and establishes relations among different properties.

A traditional way to start building thermodynamics is to identify a conserved quantity, which can be exchanged but not created. It could be matter, money, energy, etc. For most physical systems, the basic symmetry is the invariance of the fundamental laws with respect to time shifts.[3] The evolution of an isolated physical system is usually governed by the Hamiltonian (the energy written in canonical variables), whose time independence means energy conservation. In what follows, the conserved quantity of thermodynamics is called energy and denoted $E$. We wish to ascribe to the states of the system the values of $E$. First, we focus on the states independent of how they are prepared; such *equilibrium* states are completely characterized by the *static* values of observable variables.

Passing from state to state involves energy change, which generally consists of two parts: the energy change of visible degrees of freedom (which we call work) and the energy change of hidden degrees of freedom (which we call heat). To be able to measure energy changes in principle, we need adiabatic

2. In science, we strive to get the whole truth at any price. In teaching, we sell its parts at affordable prices.

3. Be careful trying to build thermodynamics for biological or social-economic systems, since generally the laws that govern them are not time invariant. For example, the metabolism of living beings changes with age, and the number of market regulations generally increases (as well as the total money mass, albeit not necessarily in our pockets).

processes, where there is no heat exchange so that all the energy changes are visible (or no under-the-table payments are made). Ascribing to every state its energy (up to an additive constant common for all states) hinges on our ability to relate any two states $A$ and $B$ by an adiabatic process, either $A \to B$ or $B \to A$, which allows us to measure the difference in their energies by the work $W$ done by the system. Now, if we encounter a process where the energy change is not equal to the work, we call the difference the heat exchange, $\delta Q$:

$$dE = \delta Q - \delta W. \tag{1.2}$$

This statement is known as the first law of thermodynamics. It is nothing but a declaration of our belief in energy conservation: if the visible energy balance does not hold, then the energy of the hidden must change. The energy is a function of the state, so we use the differential, but we use $\delta$ for heat and work, which aren't differentials of any function. Heat exchange and work depend on the path taken from $A$ to $B$; that is, they refer to particular forms of energy transfer (not energy content). Before the first law was experimentally discovered (Mayer 1842, Joule 1845), heat was believed to be a separate fluid conserved by itself.

*The basic problem of thermodynamics* is determining the equilibrium state that eventually results after all internal constraints are removed in a closed composite system. The problem is solved with the help of the extremum principle: There exists a quantity $S$ called entropy, which is a function of the parameters of the system. The values assumed by the parameters without an internal constraint maximize the entropy over the manifold of available states (Clausius 1865).

*Thermodynamic limit*   Traditionally, thermodynamics has dealt with extensive parameters whose value grows linearly with the number of degrees of freedom. Additive quantities, like the number of particles $N$, electric charge, and magnetic moment, are extensive. Energy generally is not additive; that is, the energy of a composite system is not the sum of the parts because of an interaction energy: $E(N_1) + E(N_2) \neq E(N_1 + N_2)$. To treat energy as an additive variable, we assume short-range forces of interaction acting only along the boundary and take the thermodynamic limit $V \to \infty$. Then one can neglect the interaction energy, which scales as a surface $V^{2/3} \propto N^{2/3}$, in comparison with the additive bulk terms, which scale as $V \propto N$.

In that limit, thermodynamic entropy is also an extensive variable,[4] which is a homogeneous first-order function of all the extensive parameters:

$$S(\lambda E, \lambda V, \ldots) = \lambda S(E, V, \ldots). \tag{1.3}$$

The function $S(E, V, \ldots)$, also called a *fundamental relation*, is *everything* one needs to know to solve the basic problem (and others) in thermodynamics.

Of course, (1.3) does not mean that $S(E)$ is a linear function when other parameters are fixed: $S(\lambda E, V, \ldots) \neq \lambda S(E, V, \ldots)$. On the contrary, the equilibrium curve $S(E)$ must be convex to guarantee the stability of a homogeneous state. Indeed, imagine that a system breaks spontaneously into two halves with a bit different energies, $E + \Delta$ and $E - \Delta$. For equilibration to bring back the homogeneous state, its entropy $2S(E)$ must exceed the sum of the halves: $2S(E) > S(E + \Delta) + S(E - \Delta) \approx 2S(E) + S''\Delta^2$. That requires $S'' < 0$ (the argument does not work for systems with long-range interaction where energy is nonadditive).

The figure shows the restriction imposed by thermodynamics on possible states: unconstrained equilibrium states are on the curve, while all other states lie below. Convexity guarantees that one can reach state $A$ either by maximizing entropy at a given energy or minimizing energy at a given entropy:

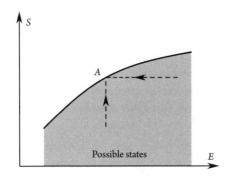

Let us complement the visual geometric picture by an analytic relation between the extrema of entropy and energy. We assume the functions $S(E, X)$ and $E(S, X)$ to be continuous differentiable for any other parameter $X$. An efficient

---

4. We shall see later that nonextensive parts of entropy are also important for studying interaction and correlations between subsystems.

way to treat partial derivatives of two functions of two variables is to organize them into a $2 \times 2$ matrix and use its determinant, called a *Jacobian*:

$$\frac{\partial(u,v)}{\partial(x,y)} \equiv \frac{\partial u}{\partial x}\frac{\partial v}{\partial y} - \frac{\partial v}{\partial x}\frac{\partial u}{\partial y}.$$

It changes sign upon any interchange of functions or variables. The partial derivative is a Jacobian:

$$\left(\frac{\partial u}{\partial x}\right)_y = \frac{\partial(u,y)}{\partial(x,y)}.$$

Then from

$$\left(\frac{\partial S}{\partial X}\right)_E = \frac{\partial(SE)}{\partial(SX)} = 0$$

follows

$$\left(\frac{\partial E}{\partial X}\right)_S = \frac{\partial(ES)}{\partial(XS)}\frac{\partial(EX)}{\partial(EX)} = -\frac{\partial(ES)}{\partial(EX)}\frac{\partial(EX)}{\partial(SX)} = -\left(\frac{\partial S}{\partial X}\right)_E\left(\frac{\partial E}{\partial S}\right)_X = 0.$$

That means that any entropy extremum is also an energy extremum. Differentiating the last relation one more time, we differentiate only the first factor since it turns into zero at equilibrium:

$$\left(\frac{\partial^2 E}{\partial X^2}\right)_S = -\left(\frac{\partial^2 S}{\partial X^2}\right)_E\left(\frac{\partial E}{\partial S}\right)_X.$$

The equilibrium is an entropy maximum; that is, $(\partial^2 S/\partial X^2)_E$ is negative. Which type of extremum has energy at equilibrium depends on the sign of $(\partial E/\partial S)_X$, which is called temperature; see (1.4) below. When the temperature is positive, as in the figure, the equilibrium is the entropy maximum at a fixed energy or the energy minimum at a fixed entropy—very much like a ball can be defined as the figure of either maximal volume $\mathcal{V}$ for a given surface area $\mathcal{A}$ or minimal area for a given volume. Such analogies create rich connections between thermodynamics and isoperimetric inequalities of the type $\mathcal{A}^d \geq d\mathcal{V}^{d-1}\mathcal{V}_0$, where $\mathcal{V}_0$ is the volume of the unit ball in $d$ dimensions.

The temperature could be negative—an example of a two-level system in section 1.4 shows that $S(E)$ could be nonmonotonic for systems with a finite phase space. Still, for every interval of a definite derivative sign, say, $(\partial E/\partial S)_X > 0$, we can solve $S = S(E, V, \ldots)$ uniquely for $E(S, V, \ldots)$, which is an equivalent fundamental relation.

Experimentally, one usually measures *changes*, thus finding derivatives. The partial derivatives of an extensive variable with respect to its arguments (which are also extensive parameters) are intensive parameters. In thermodynamics, we have only extensive and intensive variables, because we take the thermodynamic limit $N \to \infty$, $V \to \infty$, keeping $N/V$ finite. For energy, one writes

$$\frac{\partial E}{\partial S} \equiv T(S, V, N), \quad \frac{\partial E}{\partial V} \equiv -P(S, V, N) \quad \frac{\partial E}{\partial N} \equiv \mu(S, V, N), \ldots \quad (1.4)$$

These relations are called the *equations of state*, and they serve as *definitions* for temperature $T$, pressure $P$, and chemical potential $\mu$, corresponding to the respective extensive variables $S, V, N$. Our entropy is dimensionless, so $T$ is assumed to be multiplied by the Boltzmann constant, $k = 1.3 \cdot 10^{-23} J/K$, and has the same dimensionality as the energy. From (1.4), we write

$$dE(S, V, N) = \delta Q - \delta W = TdS - PdV + \mu dN. \quad (1.5)$$

The extensive parameters $V, N$ describe macroscopic (visible) degrees of freedom. Entropy is responsible for hidden degrees of freedom (i.e., heat). We shall see later that entropy is the missing information, which is thus maximal for hidden degrees of freedom in equilibrium. Temperature is the energetic price of information.

The derivatives (1.4) are taken at equilibrium, where a definite relation exists between variables, for instance, $E$ and $S$. That means that (1.5) is true only for *quasi-static processes*, i.e., such that the system is close to equilibrium at every point of the process. A process can be considered quasi-static if its typical time of change is larger than the relaxation times (which can be estimated for pressure as $L/c$, where $L$ is system size and $c$ is sound velocity, and for temperature as $L^2/\kappa$, where $\kappa$ is thermal conductivity). Finite deviations from equilibrium make $dS > \delta Q/T$ because entropy can increase without heat transfer. Only recently have we learned how to measure equilibrium quantities in fast, nonequilibrium processes, as described in section 4.4.

Let us see how the entropy maximum principle solves the basic problem. Consider two simple systems separated by a rigid wall that is impermeable to anything but heat. The whole composite system is closed; that is, $E_1 + E_2 =$ const.

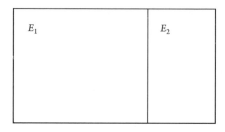

The entropy change under the energy exchange must be nonnegative:

$$dS = \frac{\partial S_1}{\partial E_1} dE_1 + \frac{\partial S_2}{\partial E_2} dE_2 = \frac{dE_1}{T_1} + \frac{dE_2}{T_2} = \left( \frac{1}{T_1} - \frac{1}{T_2} \right) dE_1 \geq 0. \quad (1.6)$$

For positive temperature, that means energy flows from the hot subsystem to the cold one: $T_1 > T_2 \Rightarrow dE_1 < 0$. We see that our definition (1.4) agrees with our intuitive notion of temperature. When equilibrium is reached, $dS = 0$, which requires $T_1 = T_2$. If the fundamental relation is known, then so is the function $T(E, V)$. Two equations, $T(E_1, V_1) = T(E_2, V_2)$ and $E_1 + E_2 =$ const, completely determine $E_1$ and $E_2$. In the same way, one can consider a movable wall and get $P_1 = P_2$ in equilibrium. If the wall allows for particle penetration, we get $\mu_1 = \mu_2$ in equilibrium.

*Example 1.1:* Consider a system that is characterized solely by its energy, which can change between zero and $E_m = N\epsilon$. The equation of state is the energy-temperature relation $E = E_m / (1 + e^{\epsilon/T})$, which tends to $E_m/2$ at $T \gg \epsilon$ and is exponentially small at $T \ll \epsilon$. In section 1.3, we identify this with a set of $N = E_m/\epsilon$ elements with two energy levels, 0 and $\epsilon$. To find the fundamental relation in the entropy representation, we integrate the equation of state:

$$\frac{1}{T} = \frac{dS}{dE} = \frac{1}{\epsilon} \ln \frac{E_m - E}{E} \Rightarrow S(E) = N \ln \frac{N}{N - E/\epsilon} + \frac{E}{\epsilon} \ln \frac{N - E/\epsilon}{E/\epsilon}.$$
$$(1.7)$$

## 1.2   Thermodynamic Potentials

Even though it is always possible to eliminate, say, $S$ from $E = E(S, V, N)$ and $T = T(S, V, N)$, getting $E = E(T, V, N)$, this *is not* a fundamental relation and it does not contain all the information. The point is, $E = E(T, V, N)$ is a partial differential equation (because $T = \partial E/\partial S$), and even if it could be integrated, the result would contain an undetermined function of $V, N$. Still, it is easier to measure temperature than entropy, so it is convenient to have a complete formalism with an intensive parameter as an operationally independent variable and an extensive parameter as a derived quantity.

Any function $Y(X)$ defines the curve on the $X, Y$ plane. We want to describe the same curve by some function of $P = \partial Y/\partial X$. It is not enough to eliminate $X$ and consider the function $Y = Y[X(P)] = Y(P)$, because such a function determines the curve $Y = Y(X)$ only up to a shift along $X$, which changes neither $Y$ nor $P$:

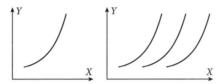

For example, the function $Y(P) = P^2/4$ corresponds to the whole family $Y = (X + C)^2$, which solves the differential equation $Y = (dY/dX)^2/4$. To pick a single function, we need to nail the curve by fixing the shift along $X$. For every $P$, we specify not $Y$ but the position $\psi(P)$, where the straight line tangent to the curve intercepts the $y$ axis: $\psi = Y - PX$:

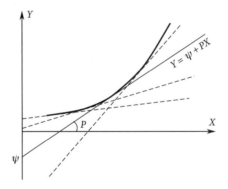

In this way, we consider the curve $Y(X)$ as the envelope of the family of the tangent lines, each characterized by the slope $P$ and the intercept $\psi$. The relation between them, $\psi(P) = Y[X(P)] - PX(P)$, completely defines the curve; here one substitutes $X(P)$ found from $P = dY(X)/dX$. The function $\psi(P)$ is called the Legendre transform of $Y(X)$. From $d\psi = -PdX - XdP + dY = -XdP$, one gets $-X = d\psi/dP$—the inverse transform is the same up to a sign: $Y = \psi + XP$.

The transform is possible when for every $X$ there is one $P$, that is, $P(X)$ is monotonic and $dP/dX = d^2Y/dX^2 \neq 0$. A sign-definite second derivative means that the function is either concave or convex. This is the second time we have met convexity, which we related above to the stability of a homogeneous state. Convexity and concavity play an important role in our story.

We can now make the Legendre transform of $E(S)$, which replaces the entropy by the temperature as an independent variable: $F = E - TS$ is called free energy. Its differential is as follows: $dF(T, V, N, \ldots) = -SdT - PdV + \mu dN + \ldots$. The counterpart to $(\partial E/\partial S)_{VN} = T$ is $(\partial F/\partial T)_{VN} = -S$. The free energy is particularly convenient for describing a system in thermal contact with a heat reservoir because the temperature is fixed, and we have one variable less to care about. The maximal work that can be done under a constant temperature (equal to that of the reservoir) is minus the differential of the free energy. Indeed, this is the work done *by the system and the thermal reservoir*. Is that work generally larger or smaller than the work done by the system alone? Let's see. That work is equal to the change in the total energy:

$$d(E + E_r) = dE + T_r dS_r = dE - T_r dS = d(E - T_r S) = d(E - TS) = dF.$$

In other words, the free energy, $F = E - TS$, is that part of the internal energy that is *free* to turn into work; the rest of the energy, $TS$, we must keep to sustain a constant temperature. The equilibrium state minimizes $F$—not absolutely, but over the manifold of states with a temperature equal to that of the reservoir. Consider, for instance, minimization of $F(T, V) = E[S(T, V), V] - TS(T, V)$ with respect to volume:

$$\left(\frac{\partial F}{\partial V}\right)_T = \left(\frac{\partial E}{\partial V}\right)_S + \left(\frac{\partial E}{\partial S} - T\right)\frac{\partial S}{\partial V} = \left(\frac{\partial E}{\partial V}\right)_S.$$

The derivatives turn into zero, and $E$ and $F$ reach extrema simultaneously. Also, in the point of an extremum, one gets $(\partial^2 E/\partial V^2)_S = (\partial^2 F/\partial V^2)_T$; i.e., both $E$ and $F$ have the same type of extremum (minimum in a positive-temperature equilibrium).

The system can reach the minimum of the free energy by minimizing energy and maximizing entropy. The former often requires creating some order in the system—for instance, orienting all spins parallel in a magnet or arranging all atoms into a regular crystal. On the contrary, increasing entropy requires disorder. Which of these tendencies wins depends on temperature, setting their relative importance. In later sections, we shall see repeatedly that

looking for a minimum of some free energy is a universal approach, from finding an equilibrium state of a physical system to designing the most optimal
algorithm of information processing.

The formal structure of thermodynamics is described in section A.1. Since
the Legendre transform is invertible, all thermodynamic potentials are equivalent and contain the same information. The choice of the potential for a given
physical situation is that of convenience: we usually take what is fixed as a
variable to diminish the number of effective variables.

The next two sections present a brief overview of the classical Boltzmann-
Gibbs statistical approach: We introduce microscopic statistical description
in the phase space and describe two principal ways (microcanonical and
canonical) to derive thermodynamics from statistics.

> **Example 1.2:** Consider a particle in the one-dimensional potential
> $U(x)$. The force $f$ one needs to apply to keep the particle in the position
> $X$ is apparently $f(X) = dU(x)/dx$ taken at $X$. Then $X(f) = dV(f)/df$,
> where $V(f)$ is minus the Legendre transform of the potential: $V(f) = Xf - U$.

## 1.3   Microcanonical Distribution

Let us consider a closed system with fixed energy $E$. Boltzmann *conjectured*
that all microstates with the same energy have equal probability (the ergodic
hypothesis). If the number of such states is $\Gamma(E)$, then the *microcanonical
probability distribution* is as follows:

$$w_a(E) = 1/\Gamma(E). \tag{1.8}$$

To link statistical physics with thermodynamics, one must define the fundamental relation, i.e., a thermodynamic potential as a function of respective
variables. For microcanonical distribution, Boltzmann in 1872 introduced
entropy as

$$S(E) = -\ln w_a(E) = \ln \Gamma(E). \tag{1.9}$$

This is one of the most important formulas in physics[5] (on a par with
$f = ma$, $E = mc^2$, and $E = \hbar\omega$).

Noninteracting subsystems are statistically independent. That means that
the statistical weight of the composite system is a product—for every state of

---

5. It is inscribed on Boltzmann's gravestone.

one subsystem, we have all the states of another. If the weight is a product, then the entropy is a sum. For interacting subsystems, this is true only for short-range forces in the thermodynamic limit $N \to \infty$.

Consider two subsystems, 1 and 2, that can exchange energy. Let's see how statistics solves the basic problem of thermodynamics (to define equilibrium) that we treated in (1.6). Assume that the indeterminacy in the energy of any subsystem $\Delta$ is much less than the total energy $E$. Alternatively, we may presume that the energy could be exchanged by portions $\Delta$. Then

$$\Gamma(E) = \sum_{i=1}^{E/\Delta} \Gamma_1(E_i)\Gamma_2(E - E_i). \tag{1.10}$$

We denote $\bar{E}_1, \bar{E}_2 = E - \bar{E}_1$ for the values that correspond to the maximal term in the sum (1.10). To find this maximum, we compute the derivative:

$$\frac{\partial \Gamma}{\partial E_i} = \frac{\partial \Gamma_1}{\partial E_i}\Gamma_2 + \frac{\partial \Gamma_2}{\partial E_i}\Gamma_1 = (\Gamma_1\Gamma_2)\left(\frac{\partial S_1}{\partial E_1} - \frac{\partial S_2}{\partial E_2}\right).$$

The extremum condition, $(\partial S_1/\partial E_1)_{\bar{E}_1} = (\partial S_2/\partial E_2)_{\bar{E}_2}$, corresponds to the thermal equilibrium where the temperatures of the subsystems are equal. The equilibrium is thus where the maximum of probability is. It is obvious that

$$\Gamma(\bar{E}_1)\Gamma(\bar{E}_2) \leq \Gamma(E) \leq \Gamma(\bar{E}_1)\Gamma(\bar{E}_2)E/\Delta \Rightarrow S(E)$$
$$= S_1(\bar{E}_1) + S_2(\bar{E}_2) + O(logN),$$

where the last term is negligible in the thermodynamic limit.

The same definition of entropy as a logarithm of the number of states is true for any system with a discrete set of states. For example, consider the set of $N$ particles (spins, neurons), each with two energy levels, 0 and $\epsilon$. If the energy of the set is $E$, then there are $L = E/\epsilon$ upper levels occupied. The statistical weight is determined by the number of ways one can choose $L$ out of $N$; that number is denoted $C_N^L$. This is our first combinatorial computation. Since we treat indistinguishable objects, let us first compute the number of permutations of $m$ things. For each of the $m$ first choices, we have $m - 1$ second choices, $m - 2$ third choices, etc. That means that the total number of permutations is $m(m - 1)(m - 2) \cdots 2 = m!$. To compute the number of ways to choose $L$ out of $N$, we need to divide the total number of permutations among $N$ by the total number of permutations among $L$ and $N - L$: $\Gamma(N, L) = C_N^L = N!/L!(N - L)!$. We can now define the entropy (i.e., find the fundamental relation): $S(E, N) = \ln \Gamma$. The entropy is symmetric about $E = N\epsilon/2$ and is zero at $E = 0, N\epsilon$, when either $L! = 1$ or $(N - L)! = 1$; that is, all the particles are in the same state. In the limit, we can use the Stirling formula, $\lim_{N\to\infty} \ln N! \approx N \ln N$. At the thermodynamic limit $N \gg 1$ and $L \gg 1$, it gives $S(E, N) \approx N \ln[N/(N - L)] + L \ln[(N - L)/L]$, which coincides with (1.7). The entropy as a function of energy is shown in the figure:

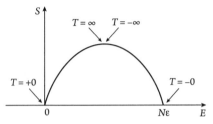

The equation of state (temperature-energy relation) is indeed $T^{-1} = \partial S/\partial E \approx \epsilon^{-1} \ln[(N - L)/L]$. We see that, when $E > N\epsilon/2$, the population of the higher level is larger than that of the lower one (inverse population as in a laser) and the temperature is negative. The negative temperature may happen only in systems with the upper limit of energy levels and simply means that, by adding energy beyond some level, we actually decrease the entropy, i.e., the number of accessible states. The example of a negative temperature is to help you disengage from the everyday notion of temperature and get used to the physicist's idea of temperature as the derivative of energy with respect to entropy. Yet it is still worth remembering the unique role played by the particular notion of temperature as mean kinetic energy of the gas molecules in the inductive development of thermodynamics.

Available (nonequilibrium) states lie below the $S(E)$ plot. The entropy maximum corresponds to the energy minimum for positive temperatures and to the energy maximum for negative temperatures. Imagine now that the system with a negative temperature is brought into contact with the thermostat (having a positive temperature). To equilibrate with the thermostat, the system needs to acquire a positive temperature. A glance at the figure shows that our system must move left, that is, give away energy (a laser generates and emits light). If this is done adiabatically slow along the equilibrium curve, the system first decreases the temperature further until it passes through minus infinity right into plus infinity and then down to positive values until it eventually reaches the thermostat's temperature. That is, negative temperatures are

actually "hotter" than positive. If you put your hand on a negative temperature system, you feel heat flowing into you. By itself, though, the system is
stable since $\partial^2 S/\partial E^2 = -N/L(N-L)\epsilon^2 < 0$ at any temperature. We stress
that there is no volume in $S(E, N)$, which means that we consider only part
of the degrees of freedom. Real particles have kinetic energy unbounded from
above and can correspond only to positive temperatures since negative temperature and infinite energy give an infinite Gibbs factor $e^{-E/T}$. Assuming
detachment between kinetic and internal (electronic, spin, etc.) degrees of
freedom is possible when their coupling is weak and only for a finite time.

The derivation of the thermodynamic fundamental relation $S(E, \ldots)$ in the
microcanonical ensemble is thus via the number of states or phase volume.

***Exercise 1.1:*** Candies and kids.

There are three candies and two systems to distribute them: system 1
contains two boys and system 2 contains three girls. Every boy and girl
can have zero, one, two, or three candies with equal probability. Kids are
distinguishable, but candies aren't.[6] What is the most probable number
of candies in system 1? What is the average number of candies in system 1? What are the most probable and average numbers of candies in
system 2?

## 1.4   Canonical Distribution and Fluctuations

Let us now discuss the statistical description, which corresponds to the thermodynamic potential of free energy, $F(T)$. Consider a system exchanging
energy with a thermostat, which can be thought of as consisting of infinitely
many copies of our system—this is the so-called canonical ensemble, characterized by $T$. Here our system can have any energy, and the question
arises, What is the probability of being in a given microstate $a$ with the
energy $E$? We derive that probability distribution (called canonical) from
the microcanonical distribution of the whole system. Since all the states of
the thermostat are equally likely to occur, the probability should be directly
proportional to the statistical weight of the thermostat $\Gamma_0(E_0 - E)$. Here we
assume $E \ll E_0$, expand (in the exponent!) $\Gamma_0(E_0 - E) = \exp[S_0(E_0 - E)] \approx$

---

6. Exchanging candies between kids leaves the system in the same state; taking candy from
one kid and giving it to another brings the system to quite a different state.

$\exp[S_0(E_0) - E/T)]$, and obtain

$$w_a(E) = Z^{-1} \exp(-E/T), \tag{1.11}$$

$$Z = \sum_a \exp(-E_a/T). \tag{1.12}$$

Note that there is no trace of the thermostat left except for the temperature. The normalization factor $Z(T, V, N)$ is a sum over all states accessible to the system and is called the partition function.

One again relates statistics and thermodynamics by defining entropy (Gibbs 1878). Recall that, for a closed system, Boltzmann defined entropy as minus the log of probability, $S = -\ln w_a$. There all probabilities were equal. Now we consider a subsystem at a fixed temperature, so that different states have different probabilities and both energy and entropy fluctuate. What should be the thermodynamic entropy: mean entropy, $-\langle \ln w_a \rangle$, or entropy at a mean energy, $-\ln w_a(E)$? They are the same! Indeed, $\ln w_a$ is linear in $E_a$ for the Gibbs distribution, so the entropy at the mean energy is the mean entropy, and we recover the standard thermodynamic relation. Comparing the mean entropy,

$$\langle S \rangle = -\langle \ln w_a \rangle = -\sum w_a \ln w_a = \sum w_a \left( E_a/T + \ln Z \right) \tag{1.13}$$
$$= E/T + \ln Z,$$

with the thermodynamic relation for it, $S = (E - F)/T$, we identify

$$F(T) = -T \ln Z(T). \tag{1.14}$$

The log of the probability of the mean energy is indeed the same as the mean log of probability:

$$S(E) = -\ln w_a(E) = -\ln \left[ \frac{\exp(-E/T)}{Z} \right] = \frac{E}{T} + \ln Z = \frac{E - F}{T}. \tag{1.15}$$

Even though the Gibbs entropy, $S = -\sum w_a \ln w_a$, is derived here for equilibrium, this definition can be used for any set of probabilities $w_a$, since it provides a useful measure of our uncertainty about the system, as we shall see in the next chapter, where entropy is a key unlocking many doors (and locking some).

The canonical equilibrium distribution corresponds to the maximum of the Gibbs entropy, $S = -\sum w_a \ln w_a$, under the condition of the given mean energy $\bar{E} = \sum w_a E_a$: Requiring $\partial(S - \beta\bar{E})/\partial w_a = 0$, we obtain (1.11). For an isolated system with a fixed energy, the entropy maximum corresponds to a uniform microcanonical distribution.

Are canonical and microcanonical statistical descriptions equivalent? Of course not. The descriptions are equivalent only when fluctuations are neglected and consideration is restricted to mean values. That takes place in thermodynamics, where the distributions produce different fundamental relations between the mean values: $S(E)$ for microcanonical, $F(T)$ for canonical. These functions are related by the Legendre transforms. Operationally, how does one check, for instance, the equivalence of canonical and microcanonical energies? One takes an isolated system at a given energy $E$, measures the derivative $\partial E/\partial S$, then puts it into the thermostat with the temperature equal to $\partial E/\partial S$; the energy now fluctuates, but the *mean* energy must be equal to $E$ (as long as the system is macroscopic and all the interactions are short-range).

As far as fluctuations are concerned, there is a natural hierarchy: microcanonical distribution neglects, and canonical distribution accounts for fluctuations in $E$. The choice of description is dictated only by convenience in thermodynamics because it treats only mean values. But if we want to describe the whole statistics of the system in a thermostat, we need to use canonical distribution, not microcanonical.

Our subsystem is macroscopic itself, so it has many ways to redistribute the energy $E$ among its degrees of freedom. In other words, it has many microscopic states corresponding to the same total energy of the subsystem. The probability for the subsystem to have a given energy is the probability of the state (1.11) times the number of states, i.e., the statistical weight of the *subsystem*:

$$W(E) = \Gamma(E)w_a(E) = \Gamma(E)Z^{-1}\exp(-E/T). \qquad (1.16)$$

The weight $\Gamma(E)$ decreases as $E \to 0$ and grows as $E \to \infty$ usually by a power law, but the exponent $\exp(-E/T)$ decays faster than any power. As a result, $W(E)$ is concentrated in a very narrow peak and the energy fluctuations around $\bar{E}$ are very small. For example, for an ideal gas, $W(E) \propto E^{3N/2}\exp(-E/T)$. To conclude, the Gibbs canonical distribution (1.11) tells us that the probability of a given microstate exponentially decays with the energy of the state, while (1.16) tells us that the probability of a given energy has a peak.

# 2

# Basics of Information Theory

This chapter presents an elementary introduction to information theory from the viewpoint of a natural scientist. It retells the story of statistical physics using a different language, which lets us see the Boltzmann and Gibbs entropies in a new light. The way of thinking here is mostly combinatoric (counting and classifying). The information viewpoint erases paradoxes and trivializes the second law of thermodynamics. It also allows us to see generality and commonality in the approaches (to partially known systems) of physicists, engineers, computer scientists, biologists, brain researchers, social scientists, market speculators, spies, and flies. We shall see how the same tools used in setting limits on thermal engines are used to set limits on communications, measurements, and learning (essentially the same phenomena). The primary mathematical tool exploits universality, which appears upon summing many independent random numbers.

The central idea developed in this chapter is that information lowers uncertainty. It can be quantified by the number of questions whose answers together eliminate the uncertainty. Suppose we are uncertain which event happens among those with a priori equal probabilities. In that case, the number of such questions is a logarithm of the number $n$ of possible outcomes, which is the Boltzmann entropy. To locate one out of $n$ equally probable objects, one needs $\log_2 n$ yes-no questions. Alternatively, one can say that the information quantifies the degree of surprise: the larger the possible number of outcomes, the more surprising any one of them is. If we know the probabilities $p_i$ of the events, then the surprise $\log_2 p_i$ is larger for the less probable ones, while the average information rate per answer is equal to the Gibbs entropy, $S = -\sum_i p_i \log_2 p_i$ bits.

But what if the answers are not entirely reliable? In other words, we have an imperfect channel whose output $A$ specifies the event (input) $B_j$ not completely but with some remaining uncertainty, characterized by the conditional entropy $S(B|A)$. The information received is then equal to $I(A, B) = S(B) - S(B|A)$ and is called mutual information.

## 2.1    Information as a Choice

Information is the resolution of uncertainty.

—CLAUDE SHANNON, 1948

We want to know in which of $n$ boxes a piece of candy is hidden. We are thus faced with a choice among $n$ equal possibilities. How much information do we need to get the candy? Let us denote the missing information by $I(n)$. Clearly, $I(1) = 0$, and we want the information to be a monotonically increasing function of $n$.[1] If we have several independent problems, then information must be additive. For example, consider each box as having $m$ compartments. To know which of the $mn$ compartments the candy is in, we first need to know which box and then which compartment inside the box: $I(nm) = I(n) + I(m)$. Now we can write (Fisher 1925, Hartley 1927, Shannon 1948)

$$I(n) = I(e) \ln n = k \ln n. \tag{2.1}$$

That information counts the number of standard questions we need to ask to specify the box. Consider yes-no questions like, "Is our candy in the right half of the set of boxes?" The answer to each question shrinks the number of boxes of interest by half. One then needs $\log_2 n$ of such questions and respective one-bit answers. If we measure information in binary choices or *bits* (abbreviation of "binary digits"), then $I(n) = \log_2 n$, that is, $k^{-1} = \ln(2)$. So the message carrying the single number of the lucky box yields the information $\log_2 n$ bits. To arrive at a destination via the road with $N$ forks, one needs $N = \log_2 2^N$ bits; via streets with $M$ intersections, one needs $M \log_2 3$ bits since there are three possible ways at each intersection.

We can easily generalize definition (2.1) for noninteger rational numbers by $I(n/l) = I(n) - I(l)$ and for all positive real numbers by considering the limits of the series and using monotonicity.

---

1. The messages "in box 1 out of 2" and "in box 1 out of 22" yield the same candy but not the same amount of information.

We used to think of information as being received through words and symbols. Essentially, it is always about which box the candy is in. If we have an alphabet with $n$ symbols, then every character we receive is a choice out of $n$ and gives the information $k \ln n$. That is, $n$ symbols are like $n$ boxes. If characters are received independently, then a message of length $N$ can potentially be one of $n^N$ possibilities, so that it yields the information $kN \ln n$. With a smaller alphabet, one needs a longer message to convey the same information. If all 26 letters of the English alphabet were used with the same frequency, then the word "love" would give information equal to $4 \log_2 26 \approx 4 \cdot 4.7 = 18.8$ bits. Here we take it for granted that the receiver has no prior knowledge of the letters (for instance, everyone who knows English can infer that there is only one four-letter word that starts with "lov," so the last letter yields zero information).

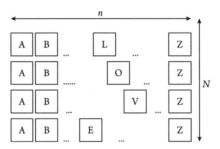

In reality, every letter on average gives even less information than $\log_2 26$ since we *know* that letters are used with different frequencies. Consider the situation when the probability $p_i$ is assigned to each letter (or box), $i = 1, \ldots, n$. It is then clear that different letters yield different degrees of surprise and different information. Let us evaluate the *average* information per symbol in a long message. To average, we consider the limit $N \to \infty$; then we know that the $i$th letter appears $Np_i$ times *in a typical sequence*: we receive the first alphabet symbol $Np_1$ times, the second symbol $Np_2$ times, etc. What we don't know and what any message of length $N$ gives us is the order in which different symbols appear. The total number of ways to place these symbols into positions (the number of different typical sequences) is equal to $N! / \Pi_i(Np_i)!$, and the information that we obtain from a string of $N$ symbols is the logarithm of that number:

$$I_N = k \ln \frac{N!}{\Pi_i(Np_i)} \approx k\left(N \ln N - \sum_i Np_i \ln Np_i\right) = -Nk \sum_i p_i \ln p_i.$$

$$(2.2)$$

The mean information per symbol coincides with the Gibbs entropy (1.13):

$$S(p_1 \ldots p_n) = \lim_{N \to \infty} I_N/N = -k \sum_{i=1}^{n} p_i \ln p_i. \qquad (2.3)$$

Alternatively, one can derive (2.3) without any mention of randomness. Consider again $n$ boxes and denote $m_i$ as the number of compartments in box number $i$. When each compartment can be chosen independently of the box it is in, the $i$th box is selected with the frequency $p_i = m_i / \sum_{i=1}^{n} m_i = m_i/M$, so that a given box is chosen more frequently if it has more compartments. The information on a specific compartment is a choice out of $M$, bringing information $k \ln M$. That information must be a sum of the information about box $I_n$ plus the information about the compartment $\ln m_i$, summed over the boxes: $k \sum_{i=1}^{n} p_i \ln m_i$. That gives the information $I_n$ about the box (letter) as the difference:

$$I_n = k \ln M - k \sum_{i=1}^{n} p_i \ln m_i = k \sum_{i=1}^{n} p_i \ln M - k \sum_{i=1}^{n} p_i \ln m_i$$

$$= -k \sum_{i=1}^{n} p_i \ln p_i = S.$$

A little more formally, one can prove that (2.3) is the only measure of uncertainty that is a continuous function of $p_i$, symmetric with respect to their permutations, and that satisfies the inductive relation

$$S(p_1, p_2, p_3 \ldots p_n) = S(p_1 + p_2, p_3 \ldots p_n) + (p_1 + p_2)S\left(\frac{p_1}{p_1 + p_2}, \frac{p_2}{p_1 + p_2}\right).$$

That relation comes from considering a subdivision: first, receive the information if one of the first two possibilities appears; second, distinguish between 1 and 2.

You probably noticed that (2.1) corresponds to the microcanonical Boltzmann entropy (1.9), giving information/entropy as a logarithm of the number of states, while (2.3) corresponds to the canonical Gibbs entropy (1.13); giving it as an average.

*Asymptotic equipartition*  Let us look at a given sequence of symbols $y_1, \ldots, y_N$ and ask, How probable is it? Can we answer this blatantly self-referential question without seeing other sequences?

Yes, we can if the sequence is long enough and we know that the symbols are independently chosen. We use the law of large numbers, which states that the sum of $N$ random numbers fast approaches $N$ times the mean value as $N$ grows. To use the law, we need to find numbers to sum. Since the symbols are independent, the probability of any sequence is the product of probabilities and the logarithm of the probability is the sum: $N^{-1} \ln P(y_1, \ldots, y_N) = N^{-1} \sum_{i=1}^{N} \ln P(y_i)$. For large $N$, it is the mean logarithm of the distribution, which is the entropy. Of which distribution? The probabilities of independent symbols depend not on the position $i$ but on which symbol from our alphabet, $y^1, y^2, \ldots, y^n$, is used. Let us denote the probabilities of different symbols as $p(y^j)$. In the limit of large $N$, we then have $N^{-1} \sum_{i=1}^{N} \ln P(y_i) = \sum_{j=1}^{n} p(y^j) \ln p(y^j)$. But how do we find $p(y^j)$? For a sufficiently long sequence, we *conjecture* that the frequencies of different symbols in our sequence give the true probabilities of these symbols. In other words, we treat the sequence as typical. Then the log of probability converges to minus $N$ times the entropy of $y$:

$$\frac{1}{N} \ln P(y_1, \ldots, y_N) \to \sum_{j=1}^{n} p(y^j) \ln p(y^j) = \langle \ln p(y) \rangle = -S(y). \quad (2.4)$$

We then state that the probability of the typical sequence decreases with $N$ exponentially: $P(y_1, \ldots, y_N) = exp[-NS(y)]$. That probability is independent of the symbols $y_1, \ldots, y_N$ so that it is the same for all typical sequences. We thus find that the best approximation for $P(y_1, \ldots, y_N)$ is a uniform (microcanonical!) distribution. Equivalently, we can state that the number of typical sequences grows with $N$ exponentially and the entropy sets the growth rate. That focus on typical sequences, which all have the same (maximal) probability, is known as asymptotic equipartition and is formulated as *"almost all events are almost equally probable."*

In physics, asymptotic equipartition is used, for instance, when we claim that the Boltzmann entropy is equivalent to the Gibbs entropy for systems whose energy is separable into independent parts in the thermodynamic limit (number of particles is an analog of a string length $N$). As we argued for equivalence of energies in section 1.4, we consider the microcanonical distribution taken at the energy equal to the mean energy of the canonical distribution (the typical set of the canonical ensemble). Then the Boltzmann $N$-particle entropy of such a microcanonical distribution is equal to the entropy of the canonical distribution in the thermodynamic limit.

Now we recognize in (2.3) the asymptotic equipartition: an $N$ string provides the information, which is the log of the number of typical strings: $I = NS$. Note that when $n \to \infty$ then (2.1) diverges while (2.3) may well be finite.

The mean information (2.3) is zero for delta distribution $p_i = \delta_{ij}$; it is generally less than the information of (2.1) and coincides with it only for equal probabilities, $p_i = 1/n$, when the entropy is maximum. Naturally, we ascribe equal probabilities when there is no extra information, i.e., in a state of maximum ignorance. In this state, a message brings maximum information per symbol; any prior knowledge can reduce the information. Mathematically, the property

$$S(1/n, \ldots, 1/n) \geq S(p_1 \ldots p_n) \tag{2.5}$$

is called convexity. It follows from the fact that the function of a single variable, $s(p) = -p \ln p$, is strictly *concave* since its second derivative, $-1/p$, is everywhere negative for positive $p$. For any concave function, the average over the set of points $p_i$ is less than or equal to the function at the average value (the so-called Jensen inequality):

$$\frac{1}{n} \sum_{i=1}^{n} s\left(p_i\right) \leq s\left(\frac{1}{n} \sum_{i=1}^{n} p_i\right). \tag{2.6}$$

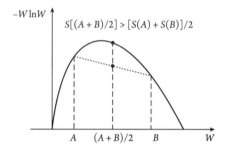

From here, one gets the entropy inequality:

$$S(p_1 \ldots p_n) = \sum_{i=1}^{n} s\left(p_i\right) \leq ns\left(\frac{1}{n} \sum_{i=1}^{n} p_i\right)$$

$$= ns\left(\frac{1}{n}\right) = S\left(\frac{1}{n}, \ldots, \frac{1}{n}\right). \tag{2.7}$$

The relations (2.6–2.7) can be proved for any concave function. The concavity condition states that the linear interpolation between two points $a, b$ lies everywhere below the function graph: $s(\lambda a + b - \lambda b) \geq \lambda s(a) + (1 - \lambda)s(b)$ for any $\lambda \in [0, 1]$; see the figure. For $\lambda = 1/2$, it corresponds to

(2.6) for $n=2$. To get from $n=2$ to arbitrary $n$, we use induction. To that end, we choose $\lambda=(n-1)/n$, $a=(n-1)^{-1}\sum_{i=1}^{n-1}p_i$, and $b=p_n$ to see that

$$s\left(\frac{1}{n}\sum_{i=1}^{n}p_i\right)=s\left(\frac{n-1}{n}(n-1)^{-1}\sum_{i=1}^{n-1}p_i+\frac{p_n}{n}\right)$$

$$\geq\frac{n-1}{n}s\left((n-1)^{-1}\sum_{i=1}^{n-1}p_i\right)+\frac{1}{n}s\left(p_n\right)$$

$$\geq\frac{1}{n}\sum_{i=1}^{n-1}s\left(p_i\right)+\frac{1}{n}s\left(p_n\right)=\frac{1}{n}\sum_{i=1}^{n}s\left(p_i\right).\qquad(2.8)$$

In the last line, we use the truth of $(2.6)$ for $n-1$ to prove it for $n$.

**Exercise 2.1:** Three squares have an average area of $100\ m^2$. The average of the lengths of their sides is $10\ m$. Use the Jensen inequality to determine the values the areas of the three squares can take.

**Exercise 2.2:** Information about precipitation.
In New York City, the probability of rain on the Fourth of July is 40%. On Thanksgiving, the probability of rain is 65%, while the probability of snow is 15%. When does the message on the presence or absence of precipitation bring more information—on Thanksgiving or on the Fourth of July?

**Exercise 2.3:** Asking the right yes-no questions.
There are two different numbers that do not exceed 100. What is the minimal number of one-bit questions we need to ask to determine both of them? How many bits does one need to find $m$ numbers not exceeding $n$?

**Exercise 2.4:** Catching counterfeit coins.
In a pile of 27 coins, there is a counterfeit coin that weighs less than the others. What is the minimum number of weighings on a balancing scale we need to isolate that coin? Describe the procedure.

*Exercise 2.5:* Deuteronomy.

Estimate the probability of the following sequence:

בראשית ברא אלוהים את השמים ואת הארץ

## 2.2  Communication Theory

Here we start treating everything as a message. After learning how much information messages bring on average, we are ready to discuss the best ways to transmit them. Communication theory is interested in two key issues—speed and reliability:

(1) How much can a message be compressed; i.e., how redundant is the information? In other words, what is the maximal transmission rate in bits per symbol?

(2) At what rate can we communicate reliably over a noisy channel; i.e., how much redundancy must be incorporated into a message to protect against errors?

Both questions concern redundancy—how unexpected is every letter of the message, on average? Entropy quantifies surprise. Let us consider a binary channel transmitting ones and zeros. Binary code is natural both for signals (present-absent) and for logic (true-false). We have seen that a communication channel transmitting independent letters transmits $- \sum_{i=a}^{z} p_i \log_2 p_i$ bits per letter. In other words, it is the average number of 0 or 1 needed to encode one letter of an alphabet.

The entropy is the mean rate of the information transfer since it is the mean growth rate of the number of typical sequences. What about the maximal rate of information transfer? Following Shannon, we answer that question statistically, which makes sense in the limit of very long messages when one can focus on typical sequences, as we did in the previous section deriving (2.2, 2.4). Consider for simplicity a message of $N$ bits, where 0 comes with probability $1 - p$ and 1 with probability $p$. To compress the message to a shorter string of letters that conveys essentially the same information, it suffices to choose a code that effectively treats the *typical* strings—those that contain $N(1 - p)$ zeros and $Np$ ones. The number of such strings is given by the binomial $C_{Np}^N$, which for large $N$ is $2^{NS(p)}$, where

$$S(p) = -p \log_2 p - (1 - p) \log_2 (1 - p).$$

The strings differ by the order of appearance of 0 and 1. To distinguish between these $2^{NS(p)}$ messages, we encode each one by a binary string with lengths starting at one and ending at $NS(p)$. For example, we encode by two one-bit words, 1 and 0, the two messages where all $Np$ ones are either at the beginning (followed by all $N(1 - p)$ zeros) or at the end (preceded by all the zeros). Then we encode by four two-bit words the messages with one hole, by eight three-bit words the messages with two holes, etc. The maximal word length $NS(p)$ is less than $N$, since $0 \leq S(p) \leq 1$ for $0 \leq p \leq 1$. In other words, to encode all $2^N$ sequences, we need words of $N$ bits, but to encode all typical sequences, we need only words up to $NS(p)$ bits. We thus achieve compression with the sole exception of the case of equal probability, where $S(1/2) = 1$. True, the code must include a few longer codewords to represent atypical messages. We then use longer and longer codewords for less and less probable sequences. In the limit of large $N$, the chance of their appearance decreases exponentially with $N$ and contribution to the transmission rate is negligible. Therefore, entropy sets both the mean and the maximal rate in the limit of long sequences. It gives the transfer rate of information when all the redundancy has been squeezed out.

> You may find it bizarre that one uses randomness in treating information communications, where one usually transfers nonrandom meaningful messages. One reason is that the encoding program does not bother to "understand" the message and treats it as random. Draining the words of meaning is necessary for devising universal communication systems.

The maximal transmission rate corresponds to the shortest mean codeword length. If we encode $n$ equally probable objects by an alphabet with $q$ symbols, the mean codeword cannot be shorter than $\log n / \log q = \log_q n$. Indeed, the number of m-letter words is $q^m$, which should not be less than $n$. For example, to encode $n = 4$ bases of the genetic code by bits $(q = 2)$, we need at least two-letter words. If we know that the objects have the probabilities $p(i)$, $i = 1, \ldots, n$, then we can use this information to shorten the mean codeword because the entropy is now lower. Shannon proved that the shortest mean length of the codeword $\ell$ is bounded by (see also section 2.8)

$$-\sum_i p(i) \log_q p(i) \leq \ell < -\sum_i p(i) \log_q p(i) + 1. \qquad (2.9)$$

Of course, only carefully chosen encoding guarantees the shortest mean codeword and the maximal rate of transmission. Designating sequences of the same length to objects with different probabilities is apparently suboptimal. Inequality (2.7) quantifies that. To shorten the mean word length and achieve signal compression in the limit of long messages, one codes frequently used objects by short sequences and infrequently used ones by lengthier combinations—lossless compressions like zip, gz, and gif work this way.

Consider a fictional creature whose DNA contains four bases, A, T, C, G, occurring with probabilities $p_i$ listed in the table:

| Symbol | $p_i$ | Code 1 | Code 2 |
|--------|-------|--------|--------|
| A | 1/2 | 00 | 0 |
| T | 1/4 | 01 | 10 |
| C | 1/8 | 10 | 110 |
| G | 1/8 | 11 | 111 |

We want a binary encoding for the four bases. As mentioned above, there are exactly four two-bit words, so one can suggest code 1, which has exactly four words and uses two bits for every base. Here the word length is two. However, it is straightforward to see that the entropy of the distribution $S = -\sum_{i=1}^{4} p_i \log_2 p_i = 7/4$ is less than two. One then may suggest a variable-length code 2. It is built in the following way. We start from the least probable C and G, which we want to have the longest codewords of the same length differing by one (last) binary digit that distinguishes between the two of them. We then can combine C and G into a single source symbol with the probability 1/4, which coincides with the probability of T. To distinguish from C, G, we code T by a two-bit word, placing 0 in the second position. The combined C, G is now encoded 11, while T is encoded 10. We then can code A by one-bit word 0 to distinguish it from the combined T, C, G.

It is straightforward now to see that code 2 uses fewer bits per base on average, namely, that its mean length of the codeword is equal to the entropy: $(1/2) \cdot 1 + (1/4) \cdot 2 + (1/4) \cdot 3 = 7/4$. The length of each codeword is exactly equal to minus the log of probability. It is an example of the so-called Huffman code, which draws a binary tree starting from its leaves: First, ascribe to the two least probable symbols the two longest codewords differing in the last digit. Second, combine these two symbols into one and repeat. The procedure ends after $n - 1$ steps, where $n$ is the size of the original alphabet. One may think that the variable-length code always requires an extra symbol (space or comma) to distinguish codewords in a continuous stream of 0 and 1. Actually, codes do not require a separating symbol if they are prefix-free, that is, no codeword can be mistaken for the beginning of another one. Such are, in particular, Huffman codes.

The most efficient code has the length of the mean codeword (the number of bits per base) equal to the entropy of the distribution, which determines the fastest mean transmission rate (i.e., the shortest mean codeword length).

To make yourself comfortable with the information brought by fractions of a bit, consider the decrease in uncertainty. One bit halves the uncertainty. For a uniform distribution, receiving one bit shrinks the uncertainty interval by the factor $2^{-1}$. Receiving $H$ bits shrinks the interval to a $2^{-H}$ fraction of its original length. Receiving a half bit shrinks the interval of possible values by the factor $2^{-1/2} \approx 0.7$.

The inequality (2.5) tells us, in particular, that using an alphabet is not optimal for the speech transmission rate as long as the probabilities of the letters are different. For example, if we use 26 letters, a space, and five punctuation marks (,.!?-), one option is to use 32 five-bit words to encode these 32 symbols (a system actually used for teletype machines). We can use variable codeword length to make the average codeword shorter than 5. Morse code uses just three symbols (dot, dash, and space) to encode any language.[2] In English, the probability of "E" is 13% and of "Q" is 0.1%, so Morse encodes "E" by a single dot and "Q" by $-\ -\ \cdot\ -$. One-letter probabilities for the written English language give the information per symbol as follows:

$$-\sum_{i=a}^{z} p_i \log_2 p_i \approx 4.11 \text{ bits,}$$

which is less than $\log_2 26 = 4.7$ bits. Language uses the same principle at the level of words: more frequently used words are generally (not always!) shorter.

**Exercise 2.6:**  Encoding by binary digits.

If we need to encode the results of independent throwing of a fair coin, we can use a one-bit encoding: 0 for heads and 1 for tails.

(a)  If we have a fair die, which is either a regular tetrahedron or a cube, how long must our binary codewords be?

2. Great contributions of Morse were the one-wire system and the simplest possible encoding (opening and closing the circuit), far superior to the multiple wires and magnetic needles of Ampere, Weber, Gauss, and many others.

(b) If we have a fair die with six sides (all having the same probability), which binary encoding could we use to provide for a transmission rate within approximately 3% of the maximal rate?

## 2.3   Redundancy in the Alphabet

*Vether it's worth goin' through so much, to learn so little, as the charity-boy said ven he got to the end of the alphabet, is a matter o' taste.*

—CHARLES DICKENS, THE PICKWICK PAPERS

The first British telegraph managed to do without C, J, Q, U, and X, which tells us that some letters can be guessed from their neighbors and, more generally, that there is a correlation between letters. Apart from one-letter probabilities, one can utilize more knowledge about the language by accounting for two-letter correlation (say, that "Q" is always followed by "U", "H" often follows "T," etc.). That further lowers the entropy.

A simple universal model with one-step correlations is called a Markov chain. It is specified by the conditional probability $p(j|i)$ that the letter $i$ is followed by $j$. For example, $p(U|Q) = 1$. The probability is normalized for every $i$: $\sum_j p(j|i) = 1$. The matrix $p_{ij} = p(j|i)$, whose elements are positive and in every column sum to unity, is called stochastic. Do the vector of probabilities $p(i)$ and the transition matrix $p_{ij}$ bring independent information? The answer is no, because the matrix $p_{ij}$ and the vector $p_i$ are not independent but are related by the condition of stationarity: $p(i) = \sum p(j)p_{ji}$, that is, $\mathbf{p} = \{p(a), \ldots p(z)\}$ is an eigenvector with the unit eigenvalue of the matrix $p_{ij}$.

The probability of an $N$ string is then the product of $N - 1$ transition probabilities times that of the initial letter. As in (2.4), minus the logarithm of the probability is a sum of uncorrelated numbers (for a Markov chain):

$$\log_2 p(i_1, \ldots, i_N) = \log_2 p(i_1) + \sum_{k=2}^{N} \log_2 p(i_{k+1}|i_k). \qquad (2.10)$$

At large $N$, the sum grows linearly with $N$ with the rate, which is the mean value of the logarithm of conditional probability, $- \sum_j p(j|i) \log_2 p(j|i) = S_i$, called the conditional entropy, $S_i$. Therefore, the number of typical sequences starting from $i$ grows with $N$ exponentially, as $2^{NS_i}$. To get the mean rate of growth for all sequences, it must be averaged over different $i$ with their probabilities $p(i)$. That way we express the language entropy via $p(i)$ and $p(j|i)$

by averaging over $i$ the entropy of the transition probability distribution:

$$S = -\sum_i p_i \sum_j p(j|i) \log_2 p(j|i). \qquad (2.11)$$

That formula defines the information rate of the Markov source. We discuss Markov chains further when describing the Google PageRank algorithm in section 4.1 and index strategies in section A.6. Let us stress that we computed not the average number of strings but the average rate of growth of this number, that is, the mean logarithm, also called the geometric mean.

One can go beyond two-letter correlations and statistically calculate the entropy of the next letter when the previous $L - 1$ letters are known (Shannon 1950). As $L$ increases, the entropy approaches the limit, which can be called the entropy of the language. Such long-range correlations and the fact that we cannot make up words bring the entropy of English down to approximately 1.4 bits per letter, *if no other information is given*. Comparing 1.4 and 4.7, we conclude that the letters in an English text are about 70% redundant— the same value one finds when asking people to guess the letters in a text one by one, which they do correctly 70% of the time. This redundancy makes data compression, error correction, and crossword puzzles possible. The famous New York City subway poster of the 1970s illustrates it:

"If u cn rd ths u cn gt a gd jb w hi pa!"

These days, it is exploited with gusto in texting, using nonstandard spelling and truncated grammar. Triple redundancy of the alphabet encoding not only serves the goal of protecting the message against errors of transmission. It could also correspond to the deeper need of our brain to obtain reinforcing of the prior guess (see section 3.4): "What I tell you three times is true!"

What is so special about the alphabet? Redundant encodings are numerous. Section A.3 explains the uniqueness of this invention and describes the frequency distribution of words and their meanings. It also explains the profound consequences of another great invention: the positional numeral system.

How redundant is the genetic code? There are four bases, which must encode 20 amino acids. There are $4^2$ two-letter words, which is not enough. The designer then must use a triplet code with $4^3 = 64$ words so that the redundancy factor is again about three. The number of ways Nature uses to encode a given amino acid is approximately proportional to its frequency of appearance.

Another example of redundancy for error protection is the NATO phonetic alphabet used by the military and pilots. To communicate through a noisy acoustic channel, full words encode letters: alpha is A, bravo is B, Charlie is C, etc.

To conclude this subsection, recall that by knowing the probability distribution, one can compute entropy, which determines the most efficient encoding rate. One can turn the tables and estimate the entropy of the data stream by looking for its most compact lossless encoding. It can be done in a one-pass (online) way, not looking at the whole data string but optimizing encoding as one processes the string from beginning to end. Several such algorithms are called adaptive codes (Lempel-Ziv, deep neural networks, etc.). These codes are also called universal since they do not require a priori knowledge of the distribution.

## 2.4  Mutual Information as a Universal Tool

In answering the first question posed in section 2.2, we have found that the entropy of the set of objects determines the minimum mean number of bits per object (word length in the binary code), which is the maximal transfer rate of the information about the objects. In this section, we turn to question (2) and find out how this rate is lowered if the transmission channel can make errors so that one cannot unambiguously restore the input $B$ from the output $A$. How much information then is lost on the way?

We continue to view everything as communications and treat measurement results $A$ as messages about the value of the quantity $B$ that we measure. We can also view storing and retrieving information as sending a message through time rather than space. We can include forecast and observation into the same scheme, asking how much information about the experimental data $B$ is contained in the theoretical predictions $A$. In all cases, $A$ is what we have and $B$ is what we want.

When the channel is noisy, the statistics of inputs $P(B)$ and outcomes $P(A)$ are generally different; we need to deal with two probability distributions and the relation between them. Treating inputs and outputs as taken out of distributions works for channels/measurements both with and without noise; in the limiting cases, the distribution can be uniform or peaked at a single value. Relating two distributions needs conditional probabilities, which we introduced in the preceding section. They lead us to relative entropy

Communication by gestures.

and mutual information, presently the most powerful and universal tools of information theory.

$$B \cdots\cdots\cdots\cdots\cdots\cdots\cdots\cdots\cdots\cdots\xrightarrow{\text{Noisy channel}} A$$

The relation between the message (measurement) $A_i$ and the event (quantity) $B_j$ is characterized by the conditional probability (of $B_j$ in the presence of $A_i$), denoted $P(B_j|A_i)$. For every $A_i$, this is a normalized probability distribution, and one can define its entropy as $S(B|A_i) = -\sum_j P(B_j|A_i) \log_2 P(B_j|A_i)$. Since we are interested in the mean quality of transmission, we average this entropy over all values of $A_j$, which defines the *conditional entropy* (Shannon called it "equivocation"):

$$S(B|A) = \sum_i P(A_i)S(B|A_i) = -\sum_{ij} P(A_i)P(B_j|A_i) \log_2 P(B_j|A_i). \quad (2.12)$$

We already encountered it in $(2.11)$ when considering correlations between subsequent terms in the sequence. If our sequence consisted of the related pairs $A_i, B_j$, like $i_k, i_{k+1}$ in the previous section, it would yield the information $(2.11, 2.12)$.

But we do not receive $B_j$. How much information about $B$ gives us knowledge of $A$? The conditional entropy between input and output measures what, on average, remains unknown about $B$ after the value of $A$ is known. The missing information was $S(B)$ before the measurement and is equal to the conditional entropy $S(B|A)$ after it. Then what the measurements give on average is their difference, called *mutual information*:

$$I(A, B) = S(B) - S(B|A) = \sum_{ij} P(A_i)P(B_j|A_i) \log_2 \left[ \frac{P(B_j|A_i)}{P(B_j)} \right]. \quad (2.13)$$

Information is a decrease in uncertainty, so mutual information must be non-negative. That means that measurements, on average, lower uncertainty by increasing the conditional probability relative to the unconditional:

$$\left\langle \log_2 \left[ \frac{P(B_j|A_i)}{P(B_j)} \right] \right\rangle \geq 0.$$

Note that we average the logarithm of the probabilities.

For example, let $B$ be a choice out of $n$ equal possibilities: $P(B) = 1/n$ and $S(B) = \log_2 n$. Assume that for every $A_i$ we can have $m$ different values of $B$ from disjoint sets, as shown on the left in the figure. Then $P(B|A) = 1/m$, $S(B|A) = \log_2 m$, and $I(A, B) = S(B) - S(B|A) = \log_2(n/m) \geq 0$, since evidently $m \leq n$. In this case, knowledge of $B$ fixes $A$, so that $S(A|B) = 0$ and $I(A, B) = S(A)$. When there is a one-to-one correspondence, $m = 1$, then $A$ tells us all we need to know about $B$.

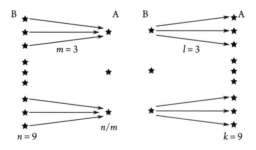

Probabilities are multiplied, and entropies are added for independent events. For correlated events, one uses conditional probabilities and entropies in what is called the chain rule:

$$P(A_i, B_j) = P(B_j|A_i)P(A_i) = P(A_i|B_j)P(B_j), \quad (2.14)$$

$$S(A, B) = S(A) + S(B|A) = S(B) + S(A|B).$$

| $S(A, B)$ | | |
|---|---|---|
| $S(A)$ | | |
| | $S(B)$ | |
| $S(A \mid B)$ | $I(A, B)$ | $S(B \mid A)$ |

This gives $I(A, B)$ in a symmetric form:

$$I(A, B) = \sum_{ij} P(A_i, B_j) \log_2 \left[ \frac{P(A_i, B_j)}{P(A_i)P(B_j)} \right] = S(B) - S(B|A)$$

$$= S(A) + S(B) - S(A, B) = S(A) - S(A|B). \qquad (2.15)$$

To illustrate the symmetry, consider the case corresponding to the $m - n$ example above: For every equally probable input $B$, we have $l$ equally probable values of $A$, whose total number is $k$, as shown on the right in the figure. In this case, $P(A|B) = 1/l$ and $S(A|B) = \log_2 l$, so that $I(A, B) = S(A) - S(A|B) = \log_2(k/l) = S(B)$ bits, similar to the $m - n$ case.

To avoid confusion, let us state the obvious: there is no symmetry between $A$ and $B$. They could be of a very different nature—one is the position of an atom and the other is the device's reading, for instance. Neither their entropies, $S(A)$ and $S(B)$, nor their conditional entropies, $S(B|A)$ and $S(A|B)$, are generally equal or even comparable. In spite of that, the degree of their correlation $I(A, B)$ is a symmetric function. If $I(A, B)$ is one bit, knowledge of the atom position shrinks by a factor of two the range of possible device readings and vice versa.

It is important to stress that measuring $A$ decreases the entropy of $B$ only on average over all values $A_i$: $S(B|A) \leq S(B)$. That follows from $P(B_j) = \sum_i P(B_j|A_i)P(A_i)$ and the convexity of the logarithm. Yet for any particular $A_i$, the entropy $S(B|A_i)$ can be either smaller or larger than $S(B)$, depending on how this measurement changes the probability distribution (see exercise 2.7). Note that $P(A_i, B_j)$ could be either larger or smaller than $P(A_i)P(B_j)$ when the pair $A_i, B_j$ are respectively correlated or anticorrelated. On average, however, the nonnegativity of the mutual information gives the so-called subadditivity of entropy:

$$S(A) + S(B) > S(A, B). \qquad (2.16)$$

When $A$ and $B$ are independent, the joint entropy is a sum, and the mutual information is zero. When $A, B$ are related deterministically, $S(A) = S(B) = S(A, B) = I(A, B)$, where $S(A) = - \sum_i P(A_i) \log_2 P(A_i)$, etc. And finally, since $P(A|A) = 1$, the mutual information of a random variable with itself is the entropy: $I(A, A) = S(A)$. So one could call entropy self-information. Another evident remark is that $I(A, B)$ exceeds neither $S(A)$ nor $S(B)$. Certainly, $A$ cannot contain more information about $B$ than about itself or than $B$ contains about itself.

> We have seen in the previous section that the mutual information between letters lowered the entropy of the language from the one-letter entropy, $-\sum_i p(i) \log p(i)$. That lowering is brought about by the knowledge of the conditional probabilities $p(j|i)$, $p(j|i, k \ldots)$, which is greater than the knowledge of $p(i)$.

## 2.5   Channel Capacity

If an imperfect channel brings about mutual information, how reliable is it? It is tempting to suggest that mutual information plays the same role for noisy channels that entropy plays for ideal channels; in particular, it sets the maximal rate of reliable communication in the limit of long messages, thus answering question (2) from section 2.2. Indeed, if there are different outputs for the same input, like in the simple $k - l$ example above, the information transfer rate is lower than for one-to-one correspondence since we need to divide our $k$ outputs into groups of $l$, distinguishing only between the groups. More formally, for each typical $N$ sequence of independently chosen $B$s, we have $[P(A|B)]^{-N} = 2^{NS(A|B)}$ possible output sequences, all of them equally likely. To get the rate of the useful information about distinguishing the inputs, we need to divide the total number of typical outputs $2^{NS(A)}$ into sets of size $2^{NS(A|B)}$ corresponding to different inputs. Therefore, we can distinguish at most $2^{NS(A)}/2^{NS(A|B)} = 2^{NI(A,B)}$ sequences of the length $N$, which sets $I(A, B)$ as the maximal rate of information transfer.

That was a rather trivial case in which inputs could be distinguished from outputs without errors so that the information transfer at that rate is reliable. But what if a single output can correspond to different inputs, like in the $m - n$ example above? There is no way now to determine every input exactly. Can we still use this imperfect channel to convey information in a way where errors can be made arbitrarily small? Yes, we can if we avoid overlapping inputs or, in other words, correctly choose the input statistics. Here we switch focus from $B$ to the communication channel or measurement procedure. Let us characterize the channel itself, maximizing $I(A, B)$ over all choices of the input statistics $P(B)$. That quantity is called Shannon's channel capacity, which quantifies the quality of communication systems in bits per symbol:

$$C = \max_{P(B)} I(A, B).$$

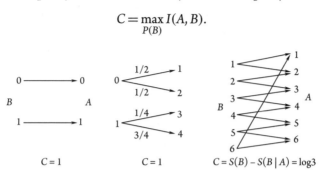

Simply put, the channel capacity is the log of the maximal number of distinguishable inputs. For example, if our channel transmits the binary input exactly (zero to zero, one to one), then the capacity is 1 bit, which is achieved by choosing $P(B=0) = P(B=1) = 1/2$; see left panel in the figure. Let us stress that if $P(0) \neq P(1)$, then the average rate is less than the capacity (one bit per symbol) despite the channel being perfect. Even if the channel has many outputs for every input out of $n$, the capacity is still $\log_2 n$, if those outputs are non-overlapping for different inputs so that the input can be determined without an error and $P(B|A) = 1$. Such a case is presented in the middle panel of the figure. In this case, the transfer rate is determined by the number of $B$ states; from the perspective of $A$ states, the rate is $S(A) - S(A|B) = 2 - 1 = 1$.

Like mutual information, the capacity deviates from $S(B)$ when the same outputs appear for different inputs, say, different groups of $m$ inputs each give the same output, so that $P(B|A) = 1/m$. In this case, one cannot achieve error-free transition for uniform $P(B)$; one needs to choose only one input symbol from each of $n/m$ groups, that is, use $P(B) = m/n$ for the symbols chosen and $P(B) = 0$ for the rest; the capacity is then indeed $C = \log_2(n/m)$ bits (right panel; $n = 6$, $m = 2$). Lowered capacity means increased redundancy, that is, a need to send more symbols to convey the same information.

Let us treat, at last, the most generic case with random errors when one cannot separate inputs/outputs into completely disjoint groups. Here one may argue that taking the limit of large $N$ does not help since the channel continues to make errors all the time. And yet Shannon showed (in the so-called noisy channel theorem) that one can both keep a finite transmission rate and make the probability of error arbitrarily small at the limit $N \to \infty$. The idea is to do error correction: send extra bits containing tests for identifying errors. To get the rate, we need to compute how many bits are devoted to error correction and how many are used to transfer the information itself. Shannon showed that it is possible to make the probability of error arbitrarily small when sending information with a finite rate $R$, if there is any correlation between output $A$ and input $B$, that is, $C > 0$. Then the probability of an error becomes $2^{-N(C-R)}$—asymptotically small in the limit of $N \to \infty$ if the rate is lower than the channel capacity. The fraction of information lost goes to zero in the limit. This (arguably most important) result of the communication theory is rather counter-intuitive: if the channel makes errors all the time, how can one decrease the error probability by treating long messages? Shannon's argument is based on typical sequences and average equipartition, that is, on the law of large numbers (by now familiar to you).

For example, if in a binary channel the probability of every single bit going wrong is $q$, then the conditional probabilities are $P(1|0) = P(0|1) = q$ and $P(1|1) = P(0|0) = 1 - q$, so that $S(A|B) = S(B|A) = S(q) = -q \log_2 q - (1 - q) \log_2(1 - q)$. The channel capacity, $C = \max_{P(B)} [S(B) - S(B|A)] = 1 - S(q)$, is achieved using the maximal entropy $S(B) = 1$ corresponding to $P(0) = P(1) = 1/2$.

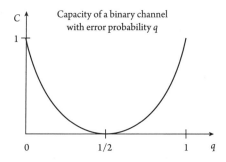

Let us now see how the capacity bounds the transmission rate from above. To correct an error, we need to specify its place. In a message of length $N$, there are $qN$ errors on average, and there are $N!/(qN)!(N - qN)! \approx 2^{NS(q)}$ ways to distribute them. We then need to devote some $m$ bits in the message not to data transmission but to error correction. Apparently, the number of possibilities provided by these extra bits, $2^m$, must exceed $2^{NS(q)}$, which means that $m > NS(q)$, and the transmission rate $R = (N - m)/N < 1 - S(q)$. In other words, flipping $qN$ bits diffuses an $N$-bit input into an "error sphere" containing $2^{NS(q)}$ typical outputs. The number of codewords is $2^{NR}$.

The error spheres of codewords do not overlap if the number strings in all spheres, $2^{NR+NS(q)}$, is less than the total number of strings, $2^N$, which gives $R < 1 - (q)$. If we wish to bound the error probability from above, we must commit to correcting more than the mean number of errors, making the transmission rate smaller than the capacity.

The conditional entropy $S(B|A)$ is often independent of the input statistics $P(B)$, as in the example above. Maximal mutual information (capacity) is then achieved for maximal $S(B)$. If no other restrictions are imposed, that corresponds to the uniform distribution $P(B)$.

If the measurement/transmission noise $\xi$ is additive, that is, the output is $A = g(B) + \xi$ with an invertible function $g$, then $S(A|B) = S(\xi)$, so that

$$I(A, B) = S(A) - S(\xi). \tag{2.17}$$

The more choices of the output that are recognizable despite the noise, the greater the capacity of the channel is. When the conditional entropy $S(A|B)$ is given, we need to choose the measurement/coding procedure to maximize the mutual information, for instance, $g(B)$ above, which maximizes the entropy of the output $S(A)$.

Mutual information also sets the limit on the data compression from $A$ to some encoding $C$ such that $S(A|C)$ is nonzero. In this case, the maximal data compression, that is, the minimal coding length in bits, is $\min I(A, C)$.

Take-home lesson: The entropy of the symbol set is the ultimate data compression rate; channel capacity is the ultimate transmission rate. Since we cannot compress below the entropy of the alphabet and cannot transfer faster than the capacity, transmission is possible only if the latter exceeds the former.

**Exercise 2.7:** Conditional entropy of criminality.

In our town, 2% of the people are criminals, and they all carry guns. In the rest of the population, only half of the people carry a gun.

(a) How much information yields a result about whether a given person is a criminal or not?
(b) How much information yields such a result if we also know in advance that the person does not carry a gun? How much

information does the result yield if we see that the person carries a gun?

(c)  How much information about a person's criminality yields knowledge of whether he/she carries a gun?

**Exercise 2.8:**  Cascade of binary channels.

Find the capacity of a cascade of $n$ consequent binary channels each with the probability of error $q$. How does the capacity decay at large $n$?

**Exercise 2.9:**  Capacity of a noisy channel.

Consider a noisy channel $X \to Y$, where both input and output can take four values. After making 128 transmissions, the frequencies were as follows:

| $Y \backslash X$ | $x_1$ | $x_2$ | $x_3$ | $x_4$ | Sum |
|---|---|---|---|---|---|
| $y_1$ | 12 | 15 | 2 | 0 | 29 |
| $y_2$ | 4 | 21 | 10 | 0 | 35 |
| $y_3$ | 0 | 10 | 21 | 4 | 35 |
| $y_4$ | 0 | 2 | 15 | 12 | 29 |
| Sum | 16 | 48 | 48 | 16 | 128 |

Compute the mutual information between the input and the output. What fraction of the output $Y$ is a signal? What would be the capacity of the channel if it were error-free?

## 2.6  Continuous Case and the Gaussian Channel

Information theory is essentially discrete since it is ultimately about counting. Moreover, the world of natural phenomena is described by digitized data both on a practical level because of finite resolution and on a fundamental level because of quantum bounds on maximal entropy in a given volume. Yet the analysis presents such a convenient mathematical tool with all the derivatives and integrals that we generalize here the definition of the Gibbs entropy (2.3) for a continuous distribution.

In a continuous case, an indeterminacy is infinite, as for the position of a point on an interval $L$. If we agree to know the position with an accuracy $\epsilon$, then the entropy of the uniform distribution is $S(B) = \log_2(L/\epsilon)$. How much information does a measurement $A$ of the point position with a precision $\Delta$ yield? The indeterminacy in the point position after the measurement is $S(B|A) = \log_2(\Delta/\epsilon)$, so that the measurement gives the information independent of $\epsilon$:

$$I(A, B) = S(B) - S(B|A) = \log \frac{L}{\Delta}. \tag{2.18}$$

We see that, even though the entropies go to infinity in the continuous limit $\epsilon \to 0$, the mutual information stays finite. That property makes the mutual information and its quantum cousin, entanglement entropy, so important in physics since they are insensitive to microscopic details and free from ultraviolet divergencies.

More generally, we define the entropy of a continuous distribution $\rho(x)$ by dividing the space of $x$ into $\epsilon$ intervals and denoting $p_i = \rho(x_i)\epsilon$. Such entropy in the limit $\epsilon \to 0$ consists of two parts:

$$-\sum_i p_i \log_2 p_i \to - \int dx \rho(x) \log_2 \rho(x) + \log_2(1/\epsilon). \tag{2.19}$$

The second term on the right is an additive constant depending on the resolution. When we are interested in the functional form of the distribution, we usually focus on the first term, which is called differential entropy:

$$S(X) = -\int dx \rho(x) \log_2 \rho(x). \tag{2.20}$$

It is not sign definite. For example, the differential entropy of the uniform distribution on the interval $L$ is $S(u, L) = L^{-1} \log_2 L$. It changes sign at $L = 1$ and $dS(u, L)/dL$ changes sign at $L = 2$ so that it does not characterize uncertainty, which we expect to grow monotonically with $L$. Yet it sheds light on the inhomogeneity of the distribution inside the interval. On a unit interval, $S(X)$ is the difference between the entropies of the coarse-grained distribution and the uniform distribution; when $\epsilon \to 0$, both diverge but their difference may stay finite. In another distinction from a discrete case, $S(X)$ is invariant with respect to shifts but not rescaling of the variables: $S(aX + b) = S(X) + \log a$.

For example, the differential entropy of the Gaussian distribution $P(\xi) = (2\pi \mathcal{N})^{-1/2} \exp[-\xi^2/2\mathcal{N}]$ is as follows:

$$S(\xi) = -\int_{-\infty}^{\infty} d\xi \, P(\xi) \log_2 P(\xi) = \frac{1}{2} \log_2 2\pi e \mathcal{N}.$$

Consider a linear noisy channel $A = B + \xi$, such that the noise is independent of $B$ and Gaussian with $\langle \xi \rangle = 0$ and $\langle \xi^2 \rangle = \mathcal{N}$. Then $P(A|B) =$

$(2\pi\mathcal{N})^{-1/2}\exp[-(A-B)^2/2\mathcal{N}]$. If in addition we have a Gaussian input signal with $P(B)=(2\pi S)^{-1/2}\exp(-B^2/2S)$, then

$$P(A)=\int dBd\xi\, P(B)P(\xi)\delta(A-B-\xi)$$

$$=[2\pi(\mathcal{N}+S)]^{-1/2}\exp[-A^2/2(\mathcal{N}+S)].$$

Now, using the chain rule, we can write

$$P(B|A)=P(A|B)P(B)/P(A)=\sqrt{\frac{\mathcal{N}+S}{2\mathcal{N}S}}\exp\left[-\frac{S+\mathcal{N}}{2\mathcal{N}S}\left(B-\frac{AS}{S+\mathcal{N}}\right)^2\right].$$

If we measure the value $A=a$, what is the best estimate for the value $B(a)=b$? It is computed using the conditional probability

$$b=\int BP(B|a)\,dB=\frac{aS}{S+\mathcal{N}}=a\frac{SNR}{1+SNR},\qquad(2.21)$$

where the signal-to-noise ratio is $SNR=S/\mathcal{N}$. The rule (2.21) makes sense: to "decode" the output of a linear detector, we use the unity factor at high $SNR$. We scale down the output at low $SNR$ since most of what we see must be noise. Note that the estimate of $b$ is linearly related to the measurement $a$, which requires linearity of the input-output relation and Gaussianity of the statistics. Let us now find the mutual information (2.17):

$$I(A,B)=S(A)-S(A|B)=S(A)-S(B+\xi|B)=S(A)-S(\xi|B)=S(A)-S(\xi)$$

$$=\frac{1}{2}\left[\log_2 2\pi e(S+\mathcal{N})-\log_2 2\pi e\mathcal{N}\right]=\frac{1}{2}\log_2(1+SNR).\quad(2.22)$$

The capacity of such a channel depends on the input statistics. One increases capacity by increasing the input signal variance, that is, the dynamic range relative to the noise. For a given input variance, the maximal mutual information (channel capacity) is achieved by a Gaussian input because the Gaussian distribution has maximal entropy for a given variance: Varying $\int dx\rho(x)(\lambda x^2+\ln\rho)$ with respect to $\rho$, we obtain $\rho(x)\propto\exp(-\lambda x^2)$. Therefore, (2.22) also determines the capacity of the Gaussian channel in bits per transmission: $C=\log_2\sqrt{(\mathcal{N}+S)/\mathcal{N}}$. That means that receiving a value $A$ allows us to distinguish between $2^C$ values. Noise effectively makes a continuous channel discrete. We elaborate on this in section 3.5. Note that the differential entropies $S(A)$ and $S(A|B)$ depend on the units used for variances and can be of either sign, while their difference and the channel capacity are positive and independent of the units.

***Exercise 2.10:*** Efficient coding of correlated Gaussian signals.

Consider two correlated signals with Gaussian statistics determined by $\langle x_1 \rangle = \langle x_2 \rangle = 0$, $\langle x_1^2 \rangle = \langle x_2^2 \rangle = 1$, and $\langle x_1 x_2 \rangle = r$. Find the most efficient encoding, $y_1(x_1, x_2)$ and $y_2(x_1, x_2)$. Remember that such encoding must maximize the data transmission rate, that is, entropy.

## 2.7 Hypothesis Testing and Bayes' Formula

... nothing but common sense reduced to calculus

—PIERRE-SIMON LAPLACE

All empirical sciences need a quantitative tool for confronting hypotheses with data. One (rational) way to do that is statistical: update prior beliefs in light of the evidence. This is done using conditional probability. For any $e$ and $h$, we have $P(e, h) = P(e|h)P(h) = P(h|e)P(e)$. If we now call $h$ the hypothesis and $e$ the evidence, we obtain the rule for updating the probability that the hypothesis is true:

$$P(h|e) = P(h)\frac{P(e|h)}{P(e)}. \tag{2.23}$$

This form of the chain rule is so important that it is named after Thomas Bayes, who first introduced it in 1763. That common-sense statement specifies how to update the probability that the hypothesis $h$ is correct after we receive the data $e$: the new (posterior) probability, $P(h|e)$, is the prior probability $P(h)$ times the quotient $P(e|h)/P(e)$, which presents the support $e$ provides for $h$. Without exaggeration, one can say that most errors made by data analysis in science and most conspiracy theories are connected to neglect or abuse of this simple formula. For example, suppose your hypothesis is the existence of a massive international conspiracy to increase the power of governments and the evidence is the COVID pandemic. In this case, $P(e|h)$ is high: a pandemic-provoking increase of state power is highly likely *given* such a conspiracy exists. Some people stop thinking right there and accept the hypothesis. They thus commit the error called inversion of the conditional since we need to evaluate not $P(e|h)$, but $P(h|e)$. Even when the former is not small, the latter could be. Indeed, absent such an event, the prior probability $P(h)$ could be vanishingly small. To overcome that smallness by a large quotient support factor, we need to evaluate total $P(e)$, that is, the probability that a pandemic happens with or without conspiracy.

If we choose between two mutually exclusive hypotheses, $h_1$ and $h_2$, the total probability of the evidence consists of two terms: $P(e) = P(e, h_1) + P(e, h_2) = P(h_1)P(e|h_1) + P(h_2)P(e|h_2)$. Then the posterior probability of the hypothesis being true is as follows:

$$P(h_1|e) = P(h_1)\frac{P(e|h_1)}{P(e)} = P(h_1)\frac{P(e|h_1)}{P(h_1)P(e|h_1) + P(h_2)P(e|h_2)}. \qquad (2.24)$$

For example, when we want to check an improbable hypothesis, $P(h_1) \ll P(h_2)$, any data changing the probability of this hypothesis won't matter much because $P(e|h_1)$ in (2.24) is multiplied by a small number, $P(h_1)$. It is better then to design an experiment or look for the data that could minimize $P(e|h_2)$ rather than maximize $P(e|h_1)$, that is, to rule out alternatives rather than support the hypothesis. This is why even good tests, with $P(e|h_1)$ close to unity and $P(e|h_2)$ small, are not very reliable at the beginning of a pandemic when $P(h_1)$ is small. The same is true for drug tests in a mostly drug-free population. Suppose that a drug test is 99% sensitive and 99% specific. That means that the test produces 99% true positive results for drug users (hypothesis $h_1$) and 99% true negative results for other people (hypothesis $h_2$). If $e$ is the positive test result, then $P(e|h_1) = 0.99$ and $P(e|h_2) = 1 - 0.99 = 0.01$. Suppose that 0.5% of people are drug users, that is, $P(h_1) = 0.005$. The probability that a randomly selected individual with a positive test is a drug user is $0.005 \cdot 0.99/(0.99 \cdot 0.005 + 0.01 \cdot 0.995) \approx 0.332$, which is less than half. The result is more sensitive to specificity approaching unity when $P(e|h_2) \to 0$ than to sensitivity. For example, taking $P(e|h_2) = 0.001$, we obtain the probability 0.83.

The choice between two (not necessarily exclusive) hypotheses is determined by the ratio of their probabilities conditioned on the data:

$$\frac{P(h_1|e)}{P(h_2|e)} = \frac{P(h_1)}{P(h_2)}\frac{P(e|h_1)}{P(e|h_2)}. \qquad (2.25)$$

Both factors on the right quantify Occam's razor, which is a preference for a simpler hypothesis. The second factor is applied to data and is mostly used by experimentalists. A more complex hypothesis, say, $h_2$, is capable of a wider variety of predictions, so it spreads its probability over the data space more thinly. If the evidence is compatible with both hypotheses (the data range is around their probability maxima, as in the figure), the simpler hypothesis generally assigns more probability to the evidence.

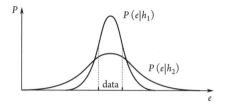

Contrary to experimentalists, theoreticians apply Occam's razor to the first factor on the right side in (2.25), choosing prior beliefs on aesthetic grounds of mathematical beauty and simplicity.

Alternatively, one can interpret higher probability as lower surprise and less information brought by the choice. That interpretation of (2.25) is sometimes called minimum description length: one should prefer the hypothesis communicating the data in fewer bits. Two subsequent messages are communicated: first, we choose the model and then communicate the data within this model. The length of the message is then $-\log_2 P(h) - \log_2 P(e|h) = -\log_2 P(e, h)$. This way, the choice of a simpler model is communicated in fewer bits, and such a model also communicates data prediction in fewer bits since a more narrow distribution has lower entropy. Technically, $P(e|h)$ is also evaluated in a two-step process, so the respective message has two parts: first, we specify the choice parameters, then communicate the data in these terms. Increasing the number of parameters, we are able to fit the data better, which shortens the error list in the data message; optimization of the respective trade-off is briefly described at the end of section 3.6.

Note the shift in the interpretation of probability brought by (2.23–2.25). The traditional sampling approach by mathematicians and gamblers treats probability as the *frequency of outcomes in repeating trials*. The Bayesian approach defines probability as a *degree of belief*; that definition allows wider applications, particularly when we cannot have repeating identical trials nor an ensemble of identical objects. For example, we have only one Earth and cannot yet restart it from the same or different initial conditions. Therefore, any estimate of the statistical significance of a global warming prediction must be based on the Bayesian approach. The approach may seem unscientific since it is dependent on prior beliefs, which can be subjective. However, by repeatedly subjecting our hypothesis to variable testing, we hope that the resulting flow in the space of probabilities will eventually come close to a fixed point independent of the starting position. Normally, only a data sequence with a clear trend of increasing probability can lead us to accept the hypothesis.

Making prior assumptions explicit is important, both computationally and conceptually. There are neither inferences nor predictions without assumptions, however uncomfortable some may feel about that (those claiming to be unbiased are usually most misleading). For example, given $5, 8, \ldots$ as two numbers of the sequence, one may put forward two hypotheses: $h_1$ predicts an arithmetic sequence $5, 8, 11, \ldots$, while $h_2$ predicts the Fibonacci sequence $5, 8, 13, \ldots$, where any number is the sum of two preceding ones. If the next number comes through the noisy channel as $12 \pm 1$, then $P(e|h_1) = P(e|h_2)$ and the choice in (2.25) is due to priors. Engineers and accountants argue that arithmetic sequences are more frequently encountered, while natural scientists point to pinecones, floral petals, and seed heads to argue for Fibonacci.

Observing our own mental processes gives us the idea of both logic and statistical inference. A Bayesian approach is used in brain research on multiple levels, from an interpretation of neural spikes and functional brain imaging to modeling sensory processing and belief propagation. One such approach is described in section 3.4.

One also uses Bayes' formula for design. For example, experimentalists measure the sensory response $A$ of an animal to the stimulus $B$, which gives $P(A|B)/P(A)$, or build a robot with the prescribed response. Then they go to the natural habitat of that animal/robot and measure the distribution of stimuli $P(B)$ (see the example at the beginning of section 3.3). After that, one obtains the conditional probability

$$P(B|A) = P(B)\frac{P(A|B)}{P(A)}, \tag{2.26}$$

which allows the animal/robot to perceive the environment and function effectively in that habitat.

## 2.8   Relative Entropy

The mutual information $I(A, B)$ measures the degree of correlation, which is essentially the difference between the true joint distribution $P(A, B)$ and the product distribution $P(A)P(B)$ of two independent quantities. As such, it is a particular case of a more general measure of difference between distributions. Let us ask the following question: How fast can a data sequence invalidate an incorrect hypothesis? If the true distribution is $p$ but our hypothetical distribution is $q$, what number $N$ of trials is sufficient to decrease the probability $P(h|e)$ by some a priori set factor? For that, we need to estimate how fast the factor $\mathcal{P} = P(e|h)/P(e)$ decreases with $N$, that is, to compute the probability of the stream of data observed given the distribution $q$. The result $i$ is observed $p_i N$ times. We *judge* the probability of that happening as $q_i^{p_i N} = e^{p_i N \ln q_i}$ times the number of sequences with those frequencies:

$$\mathcal{P} = \prod_i e^{p_i N \ln q_i} \frac{N!}{\prod_j (p_j N)!}. \tag{2.27}$$

This is how fast the probability of our hypothetical distribution being true changes with $N$, given the data set. Considering the limit of large $N$, we obtain

$$\mathcal{P} \approx \exp\left\{ N \left[ \ln N + \sum_i p_i (\ln q_i - \ln p_i N) \right] \right\}$$

$$= \exp\left[ -N \sum_i p_i \ln(p_i/q_i) \right]. \tag{2.28}$$

This is a large-deviation-type relation, like (A.7) in section A.2. The probability exponentially changes with the rate, called the *relative entropy* (Kullback-Liebler divergence):

$$D(p|q) = \sum_i p_i \ln(p_i/q_i) = \langle \ln(p/q) \rangle. \tag{2.29}$$

We need this quantity to always be nonnegative so that the probability of a not-exactly-correct hypothesis to approximate the data decreases with the number of trials. That can be shown using the simple inequality $\ln x \le x - 1$ (turning into equality only for $x = 1$):

$$-D(p|q) = \sum_i p_i \ln(q_i/p_i) \le \sum_i (q_i - p_i) = 0.$$

To prove our hypothesis wrong, the number $N$ of trials must be large enough for $ND(p|q)$ to exceed a threshold. The closer our hypothesis is to the true distribution, the larger the number of trials needed. On the other hand, when $ND(p|q)$ is below the threshold, our hypothetical distribution is just fine.

The relative entropy measures how different the hypothetical distribution $q$ is from the true distribution $p$. Note that $D(p|q)$ is not the difference between entropies (which measures the difference in uncertainties). Nor is the relative entropy a geometrical distance in the space of distributions since it does not satisfy the triangle inequality and is asymmetric: $D(p|q) \ne D(q|p)$. There is no symmetry between reality and a hypothesis. Yet $D(p|q)$ has important properties of a distance: it is nonnegative and turns into zero only when distributions coincide, that is, $p_i = q_i$ for all $i$. One possible distance between distributions is defined in exercise 2.13.

Nonnegativity and asymmetry are related for the relative entropy. If I believe that the distribution is $p_i$, then the entropy $-\sum_i p_i \ln p_i$ quantifies my average degree of surprise upon receiving the series of outcomes. But if somebody believes that the distribution is $q$, then her surprise upon the outcome $i$ is $-\ln q_i$. I *judge* her average degree of surprise to be $-\sum_i p_i \ln q_i$. That must be larger than my own degree since I naturally believe that I use the best distribution; otherwise, I'd replace it by a better option.

In particular, relative entropy quantifies how close to reality is the asymptotic equipartition estimate (2.4) of the probability of a given sequence. Assume that we have an $N$ sequence where the values/letters appear with the frequencies $q_k$, where $k = 1, \dots, d$. Then the asymptotic equipartition (the law of large numbers) suggests that the probability of that sequence is $\prod_k q_k^{Nq_k} = \exp(N \sum_k q_k \ln q_k) = \exp[-NS(q)]$. But the frequencies we observe in a finite sequence are generally somewhat different from the true probabilities $\{p_k\}$. That difference has a price so that the true probability is actually lower, which follows from the positivity of the relative entropy: $\prod_k p_k^{Nq_k} = \exp(N \sum_k q_k \ln p_k) = \exp[N \sum_k (q_k \ln q_k + q_k \ln(p_k/q_k))] = \exp\{-N[S(q) + D(q|p)]\}$.
Asymptotic equipartition, on average, overestimates the probability of a given sequence because it disregards atypical sequences, assuming that the ensemble is smaller than it really is.

If our guess is the Gibbs distribution with a given temperature, $q_i = Z^{-1}e^{-E_i/T}$, then the relative entropy is the difference of the free energies divided by that temperature:

$$D(p|q) = \ln Z + \sum_i p_i E_i/T - S(p) = -\frac{F(q)}{T} + \frac{E}{T} - S(p) = \frac{F(p) - F(q)}{T}.$$

$$(2.30)$$

The positivity of $D(p|q)$ corresponds to the known fact that the Gibbs distribution has the lowest free energy (which does not necessarily mean that it is a true distribution in every case). Therefore, one can also think of the relative entropy as a generalization of a free energy difference for a non-Gibbs $q$ distribution.

Mutual information is that particular case of the relative entropy when we compare the true joint probability $p(x_i, y_j)$ with the distribution made out of their separate measurements $q(x_i, y_j) = p(x_i)p(y_j)$, where $p(x_i) = \sum_j p(x_i, y_j)$ and $p(y_j) = \sum_i p(x_i, y_j)$:

$$D(p|q) = S(X) + S(Y) - S(X, Y) = I(X, Y) \geq 0.$$

If $i$ in $p_i$ runs from 1 to $d$, we can introduce $D(p|u) = \log_2 d - S(p)$, where $u$ is a uniform distribution. That allows us to show that both relative entropy and mutual information inherit from entropy convexity properties. You are welcome to prove that $D(p|q)$ is convex with respect to both $p$ and $q$, while $I(X, Y)$ is a concave function of $p(x)$ for fixed $p(y|x)$ and a convex function of $p(y|x)$ for fixed $p(x)$. In particular, convexity is important for ensuring that the extremum we seek is unique and lies at the boundary of allowed states.

How many different probability distributions $\{q_i\}$ (called types in information theory) exist for an $N$ sequence made out of an alphabet with $d$ symbols? The distribution $\{q_i\}$ is a $d$ vector. Since $q_i$ can take any of $N + 1$ values, $0, 1/N, \ldots, 1$, then the number of possible $d$ vectors is at most $(N + 1)^d$, which grows with $N$ only polynomially, where the alphabet size $d$ sets the power. The number of sequences grows exponentially with $N$, so that there is an exponential number of possible sequences for each type. The probability of observing a given type (empirical distribution) is determined by the relative entropy, $\mathcal{P}\{q_i\} \propto \exp[-ND(q|p)]$.

Relative entropy also measures the price of nonoptimal coding. As we discussed before, a natural way to achieve optimal coding would be to assign the length to the codeword according to the probability of the object encoded: $l_i = -\log_2 p_i$. Indeed, the information in bits about the object, $\log_2(1/p_i)$, must be exactly equal to the length of its binary encoding. This is the case with code 2 in section 2.2. For an alphabet with $d$ letters, $l_i = -\log_d p_i$. Shorter words then code the more frequently used objects, and the mean length is the entropy. The problem is that $l_i$ must all be integers, while $-\log_d p_i$ are generally not. A set of integer $l_i$ effectively corresponds to another distribution with the probabilities $q_i = d^{-l_i} / \sum_i d^{-l_i}$. Assume for simplicity that we find encoding with $\sum_i d^{-l_i} = 1$ (unity can be proved to be an upper bound for the sum). Then $l_i = -\log_d q_i$ and the mean length is $\bar{l} = \sum_i p_i l_i = -\sum_i p_i \log_d q_i = -\sum_i p_i(\log_d p_i - \log_d p_i/q_i) = S(p) + D(p|q)$, that is, larger than the optimal value $S(p)$, so that the transmission rate is lower. In particular, if one takes $l_i = \lceil \log_d(1/p_i) \rceil$ (i.e., the integer part), then one can show that $S(p) \leq \bar{l} \leq S(p) + 1$; that is, nonoptimality is at most one bit.

*Monotonicity and irreducible correlations*    If we observe fewer variables, then the relative entropy is less, a property called monotonicity:

$$D[p(x_i, y_j)|q(x_i, y_j)] \geq D[p(x_i)|q(x_j)],$$

where as usual $p(x_i) = \sum_j p(x_i, y_j)$ and $q(x_i) = \sum_j q(x_i, y_j)$. With fewer variables, we need larger $N$ to have the same confidence. In other words, information does not hurt (but only on average!). For three variables, one can define $q(x_i, y_j, z_k) = p(x_i)p(y_j, z_k)$, which neglects correlations between $X$ and the rest. What happens to $D[p(x_i, y_j, z_k)|q(x_i, y_j, z_k)]$ if we do not observe $Z$ at all? Integrating $Z$ out turns $q$ into a product. Monotonicity gives

$$D[p(x_i, y_j, z_k)|q(x_i, y_j, z_k)] = \left\langle \log \frac{p(X, Y, Z)}{p(X)p(Y, Z)} \right\rangle = S(X) + S(Y, Z)$$

$$- S(X, Y, Z) \geq D[p(x_i, y_j)|q(x_i, y_j)] = S(X) + S(Y) - S(X, Y),$$

which can be presented as the positivity of the *conditional mutual information*:

$$I(X, Z|Y) = S(X|Y) + S(Z|Y) - S(X, Z|Y) = S(X, Y) - S(Y) + S(Z, Y)$$

$$- S(Y) - S(X, Y, Z) + S(Y) = S(X, Y) + S(Y, Z) - S(Y) - S(X, Y, Z) \geq 0.$$

$$(2.31)$$

That allows one to make the next step in disentangling information encoding. The straightforward generalization of the mutual information for many objects, $I(X_1, \ldots, X_k) = \sum S(X_i) - S(X_1, \ldots, X_k)$, simply measures the total correlation. We can introduce a more sophisticated measure of correlations called the interaction (or multivariate) information, which measures the irreducible information in a set of variables beyond that which is present in any subset of those variables. For three variables, it measures the difference between the total correlation and that encoded in all pairs and is defined as follows (McGill 1954):

$$II = I(X, Z) - I(X, Z|Y) = S(X) + S(Y) + S(Z) - S(X, Y) - S(X, Z)$$

$$+ S(X, Y, Z) - S(Y, Z) = I(X, Y) + I(X, Z) + I(Y, Z) - I(X, Y, Z).$$

$$(2.32)$$

Interaction information measures the influence of a third variable on the amount of information shared between the other two and can be of either sign. When positive, it indicates that the third variable accounts for some of the correlation between the other two, so its knowledge diminishes the correlation. When negative, it indicates that the knowledge of the third variable facilitates the correlation between the other two. Alternatively, one may say that a positive $II(X, Y, Z)$ measures redundancy in the information about the third variable contained in the other two separately, while a negative one measures synergy, which is the extra information about $Y$ received by knowing $X$ and $Z$ together, instead of separately.

For example, a channel with input $X$, noise $Z$, and output $Y$ corresponds to $I(X, Z) = 0$ and $I(X, Z|Y) > 0$, that is, $II(X, Y, Z) < 0$. Indeed, once you know the output, the unknown noise and input are related. Love triangles can be either redundant or synergetic (information-wise). If $Y$ dates either $X$, both $X$, and $Z$, or none, then the dating states of $X$ and $Z$ are correlated. Knowing one tells us more about another (chooses from more possibilities) when the state of $Y$ is not known than when it is: $I(X, Z) > I(X, Z|Y)$. On the contrary, if $Y$ can date with equal probability one, another, both, or none, the states of $X$ and $Z$ are uncorrelated, but the knowledge of $Y$ induces correlation between $X$, and $Z$: if we know that $Y$ presently dates, then it is enough to know that $X$ does not to conclude that $Z$ does. Note that $II(X, Y, Z)$ is symmetric.

Capturing dependencies by using structured groupings can be generalized for an arbitrary number of variables as follows:

$$I_n = \sum_{i=1}^{n} S(X_i) - \sum_{ij} S(X_i, X_j) + \sum_{ijk} S(X_i, X_j, X_k)$$

$$- \sum_{ijkl} S(X_i, X_j, X_k, X_l) + \ldots + (-1)^{n+1} S(X_1, \ldots, X_n). \qquad (2.33)$$

Entropy, mutual information, and interaction information are the first three members of that hierarchy.

An important property of both relative entropy and all $I_n$ for $n > 1$ is that they are independent of the additive constants in the entropies, that is, of the choice of units or bin sizes. One can also define differential relative entropy, $\int dx\, \rho(x) \log[\rho(x)/\rho'(x)]$, which is invariant with respect to arbitrary (differentiable) coordinate transformations, $x \rightarrow y(x)$.

Relative entropy also allows us to generalize the second law for nonequilibrium processes. Entropy itself can either increase upon evolution toward thermal equilibrium or decrease upon evolution toward a nonequilibrium state, as demonstrated in sections 5.3 and 5.4, respectively. However, the relative entropy between the distribution and the steady-state distribution monotonously decreases with time.

**Exercise 2.11:** Interaction information.

Consider a love triangle in which $Y$ can date $X$ and $Z$. Consider the statistics of dating-not dating. Compute the entropies of the joint distribution and all the marginal distributions and the interaction information, $II = S(X) + S(Y) + S(Z) + S(X, Y, Z) - S(X, Y) - S(X, Z) - S(Y, Z)$, in the two cases.

(a) Assume that $Y$ with equal $1/3$ probabilities can be in these three states: not dating anyone, dating $X$, dating $Z$. That is, $Y$ is dating with probability $2/3$.

(b) Assume that $Y$ with equal $1/4$ probabilities can be in these four states: not dating anyone, dating $X$, dating $Z$, dating both $X$ and $Z$.

**Exercise 2.12:** Correlations between three events.

What sign is the interaction information between i) clouds, rain, and darkness, and ii) a dead car battery, a broken fuel pump, and failure to start the engine?

**Exercise 2.13:** Distance between distributions.

Consider two random quantities $X$ and $Y$ and define $\rho(X, Y) = S(X|Y) + S(Y|X)$. Apparently, $\rho(X, Y)$ is nonnegative and turns into zero if and only if $X$ and $Y$ are perfectly correlated.

(a) Prove the triangle inequality $\rho(X, Z) \leq \rho(X, Y) + \rho(Y, Z)$.

(b) Recall that a Markov chain is an ordered set of probability distributions where the next one depends only on the one immediately preceding it. In particular, the three random quantities $X \to Y \to Z$ constitute a Markov triplet if $Y$ is completely determined by $X, Z$, while $X, Z$ are independent, conditional on $Y$; that is, $I(X, Z|Y) = 0$. Find the relation between $\rho(X, Z)$ and $\rho(X, Y), \rho(Y, Z)$.

# 3

# Applications of Information Theory

This chapter gives some content to the general notions introduced thus far. Choosing from an enormous variety of applications, I tried to balance the desire to show beautiful original works and touch diverse subjects to let you recognize the same ideas in different contexts. The chapter is concerned with practicality no less than with optimality; we often sacrifice the latter for the former. The simplest and probably the most important lesson we learn here is that looking for a conditional entropy maximum is a universal approach.

## 3.1 The Whole Truth and Nothing but the Truth

So far, we have defined entropy and information via distribution. In practical applications, however, the distribution $\rho(x, t)$ is usually unknown and we need to guess it from some data. Information theory supplies a systematic way of guessing, making use of partial information, which is assumed to be given as $\langle R_j(x, t) \rangle = \int \rho(x, t) R_j(x, t) \, dx = r_j(t)$, i.e., as the ensemble averages of some dynamical quantities including normalization, $\int \rho(x, t) \, dx = 1 = -r_0$. How to get the best guess for $\rho(x, t)$, based on that information? Before, we used to find thermal equilibrium distribution looking for an entropy maximum under some condition. Now we want to treat any distribution; among the parameters that we measure could be currents, gradients, or other signs of nonequilibrium. Nevertheless, the approach is essentially the same. There are infinitely many distributions that contain *the whole truth* (i.e., are compatible with all the given information). Our distribution must also contain *nothing but the truth*; that is, it must maximize the missing information, which is the entropy $S = -\langle \ln \rho \rangle$. This provides for the widest set of possibilities for future use, compatible with the existing information (Jaynes 1957). Looking for the extremum of

$$S + \sum_j \lambda_j \langle R_j(x, t) \rangle = \int \rho(x, t) \left\{ -\ln[\rho(x, t)] + \sum_j \lambda_j R_j(x, t) \right\} dx,$$

we differentiate it with respect to $\rho(x, t)$ and obtain the equation $\ln[\rho(x, t)] = -1 + \sum_j \lambda_j R_j(x, t)$, which gives the distribution

$$\rho(x, t) = \exp\left[ \sum_j \lambda_j R_j(x, t) - 1 \right] = Z^{-1} \exp\left[ \sum_{j \geq 1} \lambda_j R_j(x, t) \right]. \quad (3.1)$$

The normalization factor,

$$Z(\lambda_i) = e^{1 - \lambda_0} = \int \exp\left[ \sum_{j \geq 1} \lambda_j R_j(x, t) \right] dx,$$

and the parameters $\lambda_i$ can be expressed via the measured quantities by using

$$\frac{\partial \ln Z}{\partial \lambda_i} = r_i. \quad (3.2)$$

The distribution (3.1) corresponds to the entropy extremum, but how do we know that it is the maximum? The positivity of relative entropy proves it. Consider any other normalized distribution $g(x)$, which satisfies the constraints $\int dx\, g(x) R_j(x) = r_j$. Then

$$\int dx\, g \ln \rho = \sum_j \lambda_i r_j - \ln Z = \int dx\, \rho \ln \rho = -S(\rho),$$

so that

$$S(\rho) - S(g) = -\int dx \left( g \ln \rho - g \ln g \right) = \int dx\, g \ln(g/\rho) = D(g|\rho) \geq 0.$$

The Gibbs distribution is (3.1), with $R_1$ being energy. When it is the kinetic energy of molecules, we have a Maxwell distribution; when it is potential energy in an external field, we have a Boltzmann distribution.

**Example 3.1:** Let us return to our "candy-in-the-box" problem (think of an impurity atom in a lattice, if you prefer physics to candy) and find the probability distribution of it being in box $j$. Different attempts give different $j$, but after many attempts we find the mean value, $\langle j \rangle = r$. The distribution giving maximal entropy for a fixed mean is exponential: $\rho(j) = Z^{-1} e^{-\lambda j}$. Using $Z = \sum_j e^{-\lambda j} = (1 - e^{-\lambda})^{-1}$ and (3.2), we find $e^{-\lambda} = r/(1+r) \equiv p$, so that we have the geometric distribution $\rho(j) = (1-p)p^j$.

Another approach is to scatter on the lattice an X-ray with wavenumber $k$. This time we find $\langle \cos(kj) \rangle = 0.3$, which gives the probability distribution

$$\rho(j) = Z^{-1}(\lambda) \exp[-\lambda \cos(kj)]$$

$$Z(\lambda) = \sum_{j=1}^{n} \exp[\lambda \cos(kj)], \quad \langle \cos(kj) \rangle = d \log Z / d\lambda = 0.3.$$

We can explicitly solve this for $k \ll 1 \ll kn$ when one can approximate the sum by the integral so that $Z(\lambda) \approx nI_0(\lambda)$, where $I_0$ is the modified Bessel function. Equation $I_0'(\lambda) = 0.3 I_0(\lambda)$ has an approximate solution of $\lambda \approx 0.63$.

The set of equations (3.2) may be self-contradictory or insufficient so that the data do not allow us to define the distribution or allow it nonuniquely. For example, consider $R_i = \int x^i \rho(x) \, dx$ for $i = 0, 1, 2, 3$. Then (3.1) cannot be normalized if $\lambda_3 \neq 0$; but having only three constants, $\lambda_0, \lambda_1, \lambda_2$, one generally cannot satisfy the four conditions. That means that we cannot reach the entropy maximum, yet one can prove that we can come arbitrarily close to the entropy of the Gaussian distribution $\ln[2\pi e(r_2 - r_1^2)]^{1/2}$.

If, however, the extremum is attainable, then the information still missing after the measurements can be computed from (3.1): $S\{r_i\} = -\sum_j \rho(j) \ln \rho(j)$. It is analogous to thermodynamic entropy as a function of (measurable) macroscopic parameters. It is clear that $S$ has a tendency to decrease whenever we add a constraint by measuring more quantities $R_i$. On the contrary, removing a constraint generally leads to entropy increase.

If we know the given information at some time $t_1$ and want to make guesses about some other time $t_2$, then our information generally gets less relevant as the distance $|t_1 - t_2|$ increases. In the particular case of guessing the distribution in the phase space, the mechanism of losing information is due to the separation of trajectories described in section 5.3. If we know that at $t_1$ the system was in some region of the phase space, the set of trajectories started at $t_1$ from this region generally fills larger and larger regions as $|t_1 - t_2|$ increases. Therefore, missing information (i.e., entropy) increases with $|t_1 - t_2|$. It works both into the future and into the past. The information approach allows one to see clearly that there is really no contradiction between the reversibility of equations of motion and the growth of entropy.

Yet there is one class of quantities where information does not age. These quantities are integrals of motion. The situation in which only integrals of motion are known is called equilibrium. When we leave the system alone, all currents dissipate and gradients diffuse. The distribution (3.1) then takes the equilibrium form, either canonical (5.3) if environment temperature is known or microcanonical if only total energy is known.

From the information point of view, the statement that systems approach thermal equilibrium is equivalent to saying that all information is forgotten except the integrals of motion. If, however, we possess the information about averages of quantities that are not integrals of motion and those averages do not coincide with their equilibrium values, then the distribution (3.1) deviates from equilibrium. Examples are fluxes and gradients.

The traditional way of thinking is operational: if we leave the system alone, it is in equilibrium; we need to act for it to deviate from equilibrium. Informational interpretation lets us see it in a new light: if we leave the system alone, our ignorance about it is maximal and so is the entropy, so that the system is in thermal equilibrium. When we act on a system in a way that gives us more knowledge of it, the entropy is lowered, and the system deviates from equilibrium.

We see that looking for the distribution that realizes the entropy extremum under given constraints is a universal, powerful tool whose applicability goes far beyond equilibrium statistical physics. It is essentially common sense expressed via simple mathematics. A beautiful example of using this approach is obtaining the statistical distribution of the ensemble of neurons. In a small window of time, a single neuron either generates an action potential or remains silent, and thus the states of a network of neurons are described naturally by binary vectors, $\sigma_i = \pm 1$. The most fundamental results of measurements are the mean spike probability for each cell, $\langle \sigma_i \rangle$, and the matrix of pairwise correlations among cells, $\langle \sigma_i \sigma_j \rangle$. One can successfully approximate the probability distribution of $\sigma_i$ by maximum entropy distribution (3.1) that is consistent with the two results of the measurement. The probability distribution of the neuron signals that maximizes entropy is as follows:

$$\rho(\{\sigma\}) = Z^{-1} \exp \left[ \sum_i h_i \sigma_i + \frac{1}{2} \sum_{i<j} J_{ij} \sigma_i \sigma_j \right], \qquad (3.3)$$

where the Lagrange multipliers $h_i, J_{ij}$ have to be chosen so that the averages, $\langle \sigma_i \rangle, \langle \sigma_i \sigma_j \rangle$ in this distribution, agree with the experiment. Such models bear the name Ising in physics, where they were first used for describing systems of spins (the model was formulated by Lenz in 1920 and solved in one dimension by his student Ising in 1925). The distribution (3.3) corresponds to the thermal equilibrium in the respective Ising model, yet it describes the brain activity, which is apparently far from thermal equilibrium (unless the person is brain dead). More in section A.4.

Looking for a conditional entropy maximum is a great way to process data. Unfortunately, we lack any guidance from first principles on which conditions to impose theoretically for describing the statistics far from thermal equilibrium. Very close to equilibrium, such a condition is often the minimum of the entropy production, but for nonequilibrium states like turbulence, we don't have any idea which or how many conditions to impose.

***Exercise 3.1:*** Distribution from information.

Consider particles having coordinates $x$ on a line: $-\infty < x < \infty$. Find the probability distribution $p(x)$ in two cases.

(a) The only information established by measurement is that the mean distance from zero is $\langle |x| \rangle = X$.

(b) The only information established by measurement is that the variance is given by $\langle x^2 \rangle = X^2$.

Which measurement provides more information on the coordinate distribution? Quantify the difference in bits.

## 3.2 Exorcising Maxwell's Demon

The demon died when a paper by Szilárd appeared, but it continues to haunt the castles of physics as a restless and lovable poltergeist.

—PETER LANDSBERG, QUOTED FROM JAMES GLEICK'S
*THE INFORMATION*

Making a measurement $R$, one changes the distribution from $\rho(x)$ to $\rho(x|R)$, which has its own *conditional* entropy:

$$S(x|R) = -\int dx dR\, \rho(R)\rho(x|R) \ln \rho(x|R) = -\int dx dR\, \rho(x,R) \ln \rho(x|R).$$

The conditional entropy quantifies our remaining ignorance about $x$ once we know $R$. Measurement decreases the entropy of the system by the mutual information (2.13, 2.15)—that is, how much information about $x$ one gains:

$$S(x) - S(x|R) = \int \rho(x, R) \ln \rho(x|R) \, dx dR - \int \rho(x) \ln \rho(x) \, dx$$

$$= \int \rho(x, R) \ln \frac{\rho(x, R)}{\rho(x)\rho(R)} \, dx dR = S(x) + S(R) - S(x, R) = \Delta I. \quad (3.4)$$

But all our measurements happen in the real world at a finite temperature. Does it matter? Yes, it determines the energy cost of measurements. Assume that our system is in contact with a thermostat having temperature $T$, which by itself does not mean that our system is in thermal equilibrium (as, for instance, a current-carrying conductor). We then can define free energy as $F(\rho) = E - TS(\rho)$. The Gibbs-Shannon entropy (2.3) and the mutual information (2.13, 3.4) can be defined for arbitrary distributions. If the measurement does not change energy (like the knowledge of which half of the box the particles are in), then the entropy decrease (3.4) increases the free energy, that is, the total work we are able to do. The first law of thermodynamics then requires that the minimal work to perform such a measurement is $F(\rho(x|R)) - F(\rho(x)) = T[S(x) - S(x|R)] = T\Delta I$.

Thermodynamics interprets $F$ as the energy we are *free* to use while keeping the temperature constant. Information theory reinterprets that in the following way: If we know everything, we can possibly use all the energy (to do work); the less we know about the system, the greater is the missing information $S$ and the less work we are able to extract. In other words, the decrease of $F = E - TS$ with the growth of $S$ measures how available energy decreases with the loss of information about the system. Maxwell had this epiphany in 1878: "Suppose our senses sharpened to such a degree that we could trace molecules as we now trace large bodies, the distinction between work and heat would vanish."

The concept of entropy as missing information allows one to understand that Maxwell's demon or any other information-processing devices do not really decrease entropy. If at the beginning one has information on the position or velocity of any molecule, then the entropy is less by this amount from the start; the entropy can only increase after using and processing the information. Consider, for instance, a particle in the box at a temperature $T$. If we know which half it is in, then the entropy (the logarithm of *available* states) is $\ln(V/2)$. That teaches us that information has thermodynamic (energetic) value at a finite temperature: by placing a piston at the midpoint of the box and allowing the particle to hit and move it, we can get the work $T\Delta S = T\ln 2$ out of the thermal energy of the particle:

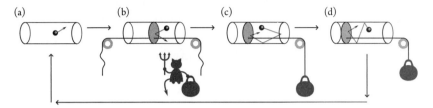

Energy conservation tells us that, to get such information, one must make a measurement whose minimum energetic cost at a fixed temperature is $W_{meas} = T\Delta S = T\ln 2$ (that was realized in 1929 by Szilard, who also introduced "bit" as a unit of information). Such work needs to be done for any entropy change by a measurement (3.4).

That guarantees that we cannot break the first law of thermodynamics. What about the second law? Our work of lifting the weight was done at the expense of the thermal energy of the system, that is, we just turned heat into work. Indeed, by hitting the moving piston, the particle loses momentum and energy, which it replenishes to $T$ by hitting the walls with that temperature provided by the environment. We can then do the measurement using this work extracted from heat. Can we break the second law by constructing a *perpetuum mobile* of the second kind, regularly using the thermal energy of the environment to do work and measuring particle position? To answer this question, we need to account for the fact that our demonic engine now includes both the working system $A$ and the measuring device $M$. For the ideal (or demonic) observer, which does not change its state upon measurements, the entropy change is the difference between the entropy of the system $S(A)$ and the entropy of the system interacting with the measuring device $S(A|M)$; that is, the *mutual information* defined in section 2.4. When there is also a change in the free energy $-\Delta F_M$ of the measuring device, the measurement work could be less than the mutual information:

$$W_{\text{meas}} \geq T\Delta S - \Delta F_M = T[S(A) - S(A|M)] - \Delta F_M. \qquad (3.5)$$

However, to make a full thermodynamic cycle, we need to return the demon's memory to the initial state. What is the energy price of *erasing* information?

Such erasure involves compression of the phase space and is irreversible. For example, to erase information about which half of the box the particle is in, we may compress the box to move the particle to one half irrespective of where it was. That compression decreases entropy and is accompanied by the heat $T \ln 2$ released from the system to the environment. If we want to keep the temperature of the system constant, we need to do exactly that amount of work compressing the box (Landauer 1961). In other words, the demon cannot get more work from using the information than we must spend erasing it to return the system to the initial state (to make a full cycle):

$$W_{\text{eras}} \geq \Delta F_M. \tag{3.6}$$

Together, the energy price of the cycle is again the mutual information:

$$W_{\text{eras}} + W_{\text{meas}} \geq T[S(A) - S(A|M)] = TI(A, M). \tag{3.7}$$

The thermodynamic energy cost of measurement and information erasure depends neither on the information content nor on the free energy difference; rather, the bound depends only on the mutual correlation between the measured system and the memory. Inequality (3.7) expresses the trade-off between the work required for erasure and that required for measurement: when one is smaller, the other one must be larger. Let us stress that information acquisition and processing have no intrinsic, irreducible thermodynamic cost, whereas the seemingly trivial act of information destruction does have a cost. The relations (3.5, 3.6, 3.7) are versions of the second law of thermodynamics, in which information content and thermodynamic variables are treated on an equal footing.

Similarly, in the original Maxwell scheme, the demon observes the molecules as they approach the shutter, allowing fast ones to pass from $A$ to $B$ and slow ones from $B$ to $A$. This is one way to use information to transfer heat from cold to hot; see also (5.13). The creation of the temperature difference with a negligible expenditure of work lowers the entropy precisely by the amount of information that the demon collected. Erasing this information also requires work.

Landauer's principle not only exorcises Maxwell's demon but also imposes the fundamental physical limit on computations. Performing standard operations independent of their history requires irreversible acts (which do not have a single-valued inverse). Any Boolean function that maps several input states onto the same output state, such as AND, NAND, OR, and XOR, is logically irreversible. When a computer performs logically irreversible operations at a finite temperature, the information is erased and heat must be generated. It is worth stressing that one cannot make this heat arbitrarily small, making the process adiabatically slow: $T \ln 2$ per bit is the minimal amount of dissipation to erase a bit at a fixed temperature.[1]

Take-home lesson: Information is physical. We can get extra work out of it, for instance, improving the efficiency of thermal engines beyond the Carnot limit. Processing information without storing an ever-increasing amount of it must be accompanied by a finite heat release at a finite temperature. Of course, any real device dissipates heat just because it works at a finite rate. Lowering that rate lowers the dissipation rate too. The message is that, no matter how slowly we process information, we cannot make the dissipation rate lower than $T \ln 2$ per bit. This is in contrast to usual thermodynamic processes where there is no information processing involved and we can make the heat release arbitrarily small, making the process slower.

## 3.3   Information Is Life

What lies at the heart of every living thing is not a fire, not warm breath, not a "spark of life." It is information.

—RICHARD DAWKINS

One may be excused for thinking that living beings consume energy and matter to survive, unless one knows that energy and matter are conserved and cannot be consumed. All the energy, absorbed by plants from sunlight and by us from food, is emitted as heat. The life-sustaining substance is entropy: we consume information and generate entropy by intercepting flows from low-entropy energy sources to high-entropy body heat. Just think how much information is processed to squeeze 500 kcal of chemical energy into 100 grams of chocolate, and you enjoy it even more. For plants, the sun is a low-entropy energy source due to its high temperature. The same is true for the whole earth, which exports into space much more entropy than it receives from the sun. Nor do we consume matter; we only make it more disordered.

---

1. In principle, any computation can be done using only reversible steps, thus eliminating the need to do work (Bennett 1973). That requires the computer to reverse all the steps after printing the answer.

What we consume has much lower entropy than what comes out of our excretory system. In other words, we decrease entropy inside and increase it outside of our bodies. Consuming information is our way to resist (temporarily) the second law of thermodynamics and survive. Atoms of our bodies are in the equilibrium state of maximal entropy and minimal free energy only postmortem.

*Genome and brain*    We have two separate systems for processing information: the genome and the brain. The genome's way of staying out of the (most probable) state of thermal equilibrium is to use replication to generate ordered (highly improbable) structures. The instructions for replication are encoded in genes. The gene is both the DNA molecule that replicates and the information that is translated to produce proteins. What are the error rates in the transmission of the genetic code? The typical energy cost of a mismatched DNA base pair is that of a hydrogen bond, which is about ten times the room temperature: $E/T \simeq 10$. If DNA molecules were in thermal equilibrium with the environment, thermal noise would cause errors, with the probability estimated from the Gibbs distribution: $e^{-E/T} \simeq e^{-10} \simeq 10^{-4}$ per base. This is deadly. A typical protein has about 300 amino acids that are encoded by about 1000 bases; we cannot have mutations in every tenth protein. Moreover, the synthesis of RNA from the DNA template and of proteins on the ribosome involves comparable energies and could cause comparable errors. That means that Nature operates in a highly nonequilibrium state, where bonding involves extra irreversible steps of removing incorrect products. This is done by molecular ratchets spending extra free energy $\Delta F$ at every step, which then fails to discriminate with the probability $e^{-\Delta F/T}$. After making $n$ steps, we lower the probability of error by the factor $e^{-n\Delta F/T}$. This way of sorting molecules is called kinetic proofreading (Hopfield 1974, Ninio 1975)

and is conceptually similar to Maxwell's demon discussed in the preceding section.

Such an effective information-processing system is the result of evolution through natural selection. The selection is not survival of the fittest. No one survives. Evolution is an increasingly efficient encoding of information about the environment in the gene pool of its inhabitants. The ultimate survivor is the information in the genes, which continues to exist long after its former carriers, individuals, and even species have gone extinct.

On another level, the nervous system maintains the body's integrity, consuming information by active inference, as described in section 3.4. The genome's method of information processing is clearly digital; what about the brain? Since neurons often either do or do not fire a standard pulse, it may seem that information is encoded in binary digits. Indeed, written language and many similar tasks are clearly handled by processing digital information. However, there are reasons to believe that the brain is also an analog device; for instance, encoding information in the frequency of pulses, which could be varied continually.

Genetic code and human language are the only known natural digital mechanisms of information storage and transmission with potentially unlimited heredity; i.e., they comprise an indefinitely large number of structures that can replicate. The genome uses a homologous pairing of four bases; just a million bases (typical for any bacteria) provide $4^{10^6}$ possible structures to replicate, exceeding the number of atoms in the universe. The modular character of the genome means that all those structures are potentially different. Similarly, in language, just adding a comma changes the meaning of the whole message: from "eats shoots and leaves" to "eats, shoots, and leaves." At the level of grammar, the language uses recursion for generating new meanings: "This is the cat that killed the rat that ate the malt that lay in the house that Jack built." Another similarity is that memes replicate like genes.[2] The combination of an infinite range of messages with a high-fidelity transmission mechanism is unique for genetic code and language.

If such an elementary act of life as information processing (say, thought) generates $\Delta S$, we can now ask about its energy price. Similar to our treatment of thermal engine efficiency $(1.1)$, we take $Q$ from the reservoir with $T_1$ and deliver $Q - W$ to the environment with $T_2$. Then $\Delta S = S_2 - S_1 =$

---

2. *Meme* is defined as a unit of cultural inheritance.

$(Q - W)/T_2 - Q/T_1$ and the energy price is as follows:

$$Q = \frac{T_2 \Delta S + W}{1 - T_2/T_1}.$$

When $T_1 \to T_2$, the information processing is getting prohibitively ineffective, just like the thermal engine. In the other limit, $T_1 \gg T_2$, one can neglect the entropy change on the source, and we have $Q = T_2 \Delta S + W$. Hot sun is an energy source of a very low entropy.

How many bits do we consume per second? Let us estimate our rate of information processing and entropy production. An average lazy human being dissipates about $W = 200$ watts of power at $T = 300$ K. Since the Boltzmann constant is $k = 1.38 \times 10^{-23}$, that gives about $W/kT \simeq 10^{23}$ bits per second. The amount of information processed per unit of subjective time (per thought) is about the same, assuming that each moment of consciousness lasts about a second (Dyson 1979).

We thus conclude that living beings obtain information from the sources of energy, matter, and, last but not least, sensors.

*Processing sensory information*    Do the Gibbs entropy and the mutual information have any quantitative relation to the way we react to signals? Yes, they do! When one must react differently to different stimuli, the average choice reaction time $T$ is linearly proportional to the entropy of the distribution of stimuli (Hick 1952, Hyman 1953). The greater the uncertainty, the longer it takes to recognize the event. For example, when one needs to name the number that appears randomly on a screen, the average response time grows logarithmically with the size of the set. Logarithmic dependence on the set size means that the decision is made by a subdividing strategy. Similarly, the time to find an item in an ordered menu grows logarithmically with the menu length; the time grows linearly when the menu is disordered.

When the number of elements stays constant but the frequencies of their appearances are made unequal, thus lowering entropy, the average response time decreases proportionally. Even more remarkably, when experimentalists introduce a correlation between subsequent stimuli, the response time goes down in proportion to the conditional entropy, which is less than unconditional.

One can turn the tables and prescribe the reaction time. As this time gets shorter, we make more and more errors in naming the objects. How to quantify that? We need to compute the mutual information between the input (the number $i$ on a screen) and the output (our name $j$ for it). Making more errors means lower mutual information. Experimentally, one measures the joint probability $p(i,j)$ from which one obtains the marginal probabilities $p(i) = \sum_j p(i,j)$, $p(j) = \sum_i p(i,j)$ and conditional probability $p(j|i) = p(i,j)/p(i)$. One then computes $S(j) = -\sum_i p(j) \log p(j)$, $S(j|i) = -\sum_{ij} p(i)p(j|i) \log p(j|i)$ and $I(i,j) = S(j) - S(j|i)$. The mutual information is found to be linearly proportional to the reaction time prescribed. Is the invention (or discovery?) of the Boltzmann and Gibbs entropies and the mutual information related to the fact that our brain actually employs them?

Since the time of processing is proportional to the amount of information, one can conclude that what characterizes the system is the average amount of information processed per unit time, that is, the rate. The next example presents a strategy for processing stimuli, where the system maximizes the information transfer rate by keeping it uniform through the dynamic range of the signal (such strategies are sometimes called the infomax principle).

*Maximizing capacity*   Imagine yourself on day five of Creation designing the response function for the sensory system of a living being. Technically, the problem is to choose thresholds for switching to the next level of response, or equivalently, to choose the function of the input for which we take equidistant thresholds. Suppose that we divide the whole perceivable (finite) interval of signals into three regions, encoding them as weak $(1, 2)$, medium $(2, 3)$, and strong $(3, 4)$:

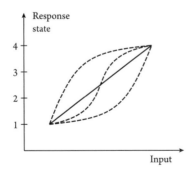

For given value intervals of input and response, should we take the solid line of linear proportionality between response and stimulus? Or choose the lowest curve that treats even medium-intensity inputs as weak and amplifies the difference in high-intensity signals? The choice may depend on the significance of different intervals for survival. For example, the upper curve was actually chosen (on day six) for the auditory system of animals and humans: our ear senses loudness as the logarithm of the intensity, which amplifies differences between weak sounds and damps strong ones. That way we better hear the whisper of someone close or the sound of a distant creek and aren't that frightened by thunder.

If, however, all the input amplitudes are of comparable significance, then the goal could be to maximize the mean information transfer rate (capacity) at the level of a single neuron/channel. In such a case, the response curve (encoding) must be designed by the Creator together with the probability distribution of stimuli. That was demonstrated in one of the first applications of information theory to real data in biology, namely, to processing of visual signals (Laughlin 1981). It was conjectured that maximal-capacity encoding must use all response levels with the same frequency, which requires that the response function is the integral of the probability distribution of the input signals (see figure). First-order interneurons of the insect eye were found to code contrast rather than absolute light intensity. Subjecting the fly in the lab to different contrasts $x$, the response function $y = g(x)$ was measured from the fly neurons; the probability density of inputs, $\rho(x)$, was measured across its natural habitat (woodlands and lakeside) using a detector that scanned horizontally, like a turning fly.

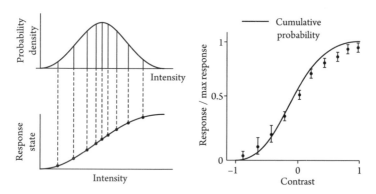

We can now explain the relation between the response and the cumulative probability by noting that the representation with the maximal capacity corresponds to the maximum of the mutual information between input and output: $I(x, y) = S(y) - S(y|x)$. Since we consider a one-to-one relation $y = g(x)$, that is, an error-free transmission, then the conditional entropy $S(y|x)$ is zero. Therefore, according to section 2.4, we need to maximize the entropy of the output, assuming that the input statistics $\rho(x)$ are given. Absent any extra constraints except normalization, the entropy for a distribution on a finite interval is maximal when $\rho(y)$ is constant. Indeed, since $\rho(y)dy = \rho(x)dx = \rho(x)dydx/dy = \rho(x)dy/g'(x)$, then

$$S(y) = -\int \rho(y) \ln[\rho(y)] \, dy = -\int \rho(x) \ln[\rho(x)/g'(x)] \, dx, \quad (3.8)$$

$$\frac{\delta S(y)}{\delta g} = \frac{\partial}{\partial x} \frac{\rho}{g'(x)} = 0 \quad \Rightarrow \quad g'(x) = C\rho(x),$$

as in the figure. In other words, we choose equal bins for the variable whose probability is flat. Since the probability $\rho(x)$ is positive, the response function $y = g(x)$ is always monotonic, i.e., invertible. Note that our choice of response function is an exact analog of efficient encoding, using longer codewords for less frequently used letters. Analogous to code 2 in section 2.2, we combine signals into intervals with the same probability. In that way, we utilize only the probability distribution of different signal levels, similar to language encoding based on frequency of letters (and not, say, their mutual correlations). We have also applied quasi-static approximation, neglecting dynamics and relating instantaneous values of $x$ and $y$. Allow yourself to be impressed by the agreement of theory and experiment—there are no fitting parameters. The same approach also works well for biochemical and genetic input-output relations. For example, the dependence of a gene expression on the level of a transcription factor is dictated by the statistics of the latter. That also works when the conditional entropy $S(y|x)$ is nonzero but independent of the form of the response function $y = g(x)$. See more details in section A.5.

For particular types of signals, practicality may favor nonoptimal but simple schemes like amplitude and frequency modulation (both are generally nonoptimal but computationally feasible and practical). Even in such cases, the choice is dictated by the information-theory analysis of the efficiency. For example, a neuron either fires a standard pulse (action potential) or stays silent, which makes it natural to assume that the information is encoded as binary digits (zero or one) in discrete equal time intervals. One can also imagine that the information is encoded by the time delays between subsequent pulses. Since time is continuous, this is more of an analog computation. In the engineer's language, the former method of encoding is a limiting case of amplitude modulation, while the latter case is one of frequency modulation. The maximal rate of information transmission in the former case is dependent only on the minimal time delay between the pulses determined by the neuron recovery time. On the other hand, in the latter case, the rate depends on both the minimal error of timing measurement and admissible maximal time between pulses. In reality, brain activity "depends in one way or another on all the information-bearing parameters of an impulse—both on its presence or absence as a binary digit and on its precise timing" (MacKay and McCulloch 1952).

## 3.4    Whom to Believe: My Eyes or Myself?

How is sensory information processed and how does it determine behavior? An ambitious application of information theory is an attempt to understand and quantify sentient behavior. One idea going back to Helmholtz is to view "perception as an unconscious inference." There is evidence that the perception of our brain is inferential, that is, based on prediction and hypothesis testing. Among other things, this is manifested by the long-known phenomenon of binocular rivalry, which occurs when different pictures are presented to our two eyes. Rather than perceiving a stable, single amalgam of the two stimuli, we experience alternations as the two stimuli compete for perceptual dominance, which can be influenced by priming. Another piece of evidence is the recently established fact that signals between the brain and sensory organs travel in both directions simultaneously.

Perception is thus treated not as a bottom-up encoding of sensory states $Y$ into internal neuronal representations of the environmental states $X$, but as a combination of top-down prior expectations with bottom-up sensory signals. The combined bottom-up-top-down approach makes sense from evolutional and developmental perspectives. The bottom-up approach, taken alone, premises some entity that processes the sensory inputs $Y$ into a picture of the world $P(X|Y)$. Yet where did that entity come from? Imagine the brain as a bunch of neurons in a black box receiving electrical signals, which do not carry with them labels "from the retina," "from the liver," "from your grandmother," etc. The best we can do is to send out signals that help us survive. Since we have managed to survive up to this point, then the right survival strategy is a continuation, in which we try to receive more or less the same signals as before.

In this spirit, we might describe perception as hypothesis testing within the Bayes approach, introduced in section 2.7. The mechanics of the sensory system determine $P(Y|X)$, which is the conditional probability of sensory input for a given state of the environment. In the example of the fly eye from section 3.3, $x$ is a contrast in light intensity and $y$ is the neuron signal. Upon receiving the particular input $y$, the simplest inference about the environment is that of maximal likelihood: taking the value $x$ that maximizes $P(y|x)$. However, to make a decision or action based on inference, we need a measure of confidence in the result. That means that our inference must be probabilistic, obtaining the whole posterior probability distribution $P(X|Y)$—sharp distribution gives a high and flat distribution low confidence. To obtain the posterior distribution, we need a prior distribution $P(X)$ and Bayes' formula (2.23):

$$P(X|Y) = P(Y|X)P(X)/P(Y). \tag{3.9}$$

Leaving aside for a while the normalizing factor $P(Y)$, we thus presume that the mind has a so-called generative model, represented by the joint distribution $P(X, Y) = P(Y|X)P(X)$. Exact computation by (3.9) can be impossible or impractical, for instance, due to the necessity of having to average over many hidden states and variables. It is natural to assume that the brain uses a variational approach based on optimizing some tractable proxy. The first thing to account for is the degree of surprise or necessary change, characterized by the relative entropy between prior and posterior distributions. Averaged over all $X$ and $Y$, it is simply the mutual information, that is, the average information brought by sensory inputs:

$$D[P(X|Y)|P(X)] = \sum_{Y} P(Y) \sum_{X} P(X|Y) \log[P(X|Y)/P(X)] = I(X, Y).$$

For perception, however, we need to evaluate the change at a given $y$. Changing beliefs and updating expectations entails a cognitive metabolic cost, as we know all too well. More important and probably related, expected states are preferred for survival (fish expect to stay in the water), while surprises are to be avoided. A generative model must be strongly biased toward a narrow interval of parameters guaranteeing survival. This natural tendency to minimize change conflicts with the necessity to accommodate data. Whenever we encounter a trade-off, free energy negotiates it. The working hypothesis is that, for a given $y$, the brain looks for the posterior distribution $Q(X)$ that minimizes the following free energy, which is a function of $y$ and a functional of $Q(X)$:[3]

---

3. A function gives a value for every value; a functional gives a value for every function.

$$F[Q(x), y] = \sum_x Q(x) \log \frac{Q(x)}{P(x, y)} = -\sum_x Q(x) \log P(x, y) - S(Q)$$

$$= D[Q(x)|P(x)] - \sum_x Q \log P(y|x) = \sum_x Q(x) \left[ \log \frac{Q(x)}{P(x)} - \log P(y|x) \right]$$

$$= D[Q(x)|P(x|y)] - \log P(y). \tag{3.10}$$

As is clear from the beginning of the first line, it measures the mismatch between the internal generative model $P(x, y)$ and the current observation. The three lines suggest three different operational strategies according to the three different interpretations of the same quantity.

Minimization of the first line thus requires the trade-off between the data-imposed "truth" and "nothing but the truth" maximization of the entropy $S(Q)$. Indeed, the logarithm of probability, as we have seen in section 3.1, is essentially the set of our prior data. Therefore, the first term on the right represents the "truth" imposed by the data—both the prior data that formed $P(x, y)$ and the given input value of $y$.

The second line describes the trade-off between inertia and the force of data: the first term on the right is the degree of change, while the second term quantifies the accuracy of data representation—$Q(x)$ must give more weight to that $x$, which provides for higher probability to observe $y$ according to $P(y|x)$, which is given.

The third line in (3.10) does not describe any trade-off but shows that the free energy is bounded from below by the sensory surprise $- \log P(y)$. Only when our variational $Q(x)$ is equal to the exact $P(x|y)$, does the free energy reach its global minimum. That suggests that perceptual inference, that is, computing $Q(x)$, is not the only way to minimize $F(Q, y)$; another way is to change the sensory data $y$. Changing input requires action: one can switch the channel or look the other way rather than change beliefs.

That brings us to the active inference approach, which puts action into perception (Parr, Pezzulo, Friston 2022). The *assumption* is that living beings survive by adapting action-perception loops to their environment. That means that every sensory input is not obtained passively, but is predicted by the brain and is solicited by an action intended for the predicted input. A mismatch between predicted and actually received sensory input leads to updating the predictive (generative) model, which then triggers new action leading to new sensory observations better corresponding to expectations. Perception and action are complementary ways to diminish the mismatch. Perception changes your mind, replacing prior beliefs with posterior ones, while action changes the world to make it more compatible with the beliefs. Surprise minimization by active inference is our way to survive.

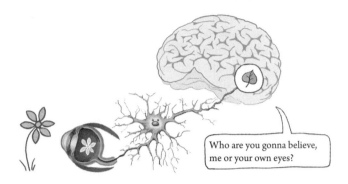

In particular, our perception of objects is very much determined by the generative model with its prediction of how actions change sensory input (encoded in the conditional probability of *what could have happened*). Even with one eye closed, we distinguish a three-dimensional object from its two-dimensional picture despite receiving identical visual signals. The reason is that our brain knows that moving our head will reveal the new parts of the image in the former case but not in the latter.

While still highly hypothetical, this theory finds some empirical support in measurements of the connectivity and activity of neural networks. For example, some connectivity patterns in a motor cortex support the idea of a motor command as a prediction, such that the prediction errors related to body position and motion can be resolved by reflexes without belief updating. Simply speaking, the brain can infer the positions of body parts without receiving outside signals. The analysis of the experimental data on brain activity is facilitated by the asymmetry between descending signals carrying expectations and ascending signals bringing prediction errors—the latter involves nonlinear operations generating higher frequencies, which is measurable. Last but not least—playing tennis would be impossible if our brain just reacted to visual stimuli, since the time between light hitting the retina and the brain receiving a signal is in excess of 150 msec. The active inference approach is also useful in building models for analyzing data from behavioral experiments and disease processes and drawing inferences about inferences. When top-down signals totally dominate, one has hallucinations; what is considered normal perception could then be called "controlled hallucination."

The internal generative model encodes the world from the perspective of the body's needs, not in some "objective" way, which would be a waste of metabolic resources. Physiological state changes constantly, shifting the focus to different sensory inputs. The neurons then are not expected to lie dormant waiting for sensory input. In particular, emotions must play an important role in choosing the focus. Within the active inference approach, emotions are treated not solely as fixed universal patterns of brain and body inherited from animal ancestors and triggered by sensory inputs. One may consider emotions as constructed and learned patterns of prediction and reaction amenable to significant variability and plasticity. I find it inspiring that our work with averaging logarithms and finding conditional optima could one day have direct moral implications.

I mention in passing the suggestions to use relative information and mutual entropy for the more ambitious task of quantifying consciousness, understood as processing information from different channels in an integrated way, irreducible to processing information in the channels separately. Such an approach is known as integrated information theory (Tononi 2008).

I think that poetry and music appeal to our ever-predicting mind by creating expectations (using rhythm or melody) and then partially fulfilling and partially breaking them. An optimal mixture of expected and surprising is what makes for great art, which still waits for its free energy analysis.

## 3.5   Rate Distortion and the Information Bottleneck

There's no sense in being precise when you don't even know what you are talking about.

—JOHN VON NEUMANN

When we transfer information, we look for the maximal transfer rate and thus define channel capacity as the maximal mutual information between input and output. But when we encode information, we may be looking for the opposite: What is the *minimal* number of bits, sufficient to encode the data with a given accuracy?

For example, encoding a real number requires an infinite number of bits. Representation of a continuous input $B$ by a finite discrete output encoding $A$ generally leads to some distortion, which we characterize by the real function $d(A, B)$. How large is the mean distortion, $\mathcal{D} = \sum_{ij} P(A_i, B_j) d(A_i, B_j)$, for given statistics of $B$ and the encoding $A$ with $R$ bits and $2^R$ values? It depends on the choice of the distortion function, which specifies the most important properties of the signal $B$. For Gaussian statistics (which is completely determined by the variance), one chooses the squared error function, $d(A, B) = (A - B)^2$. We first learn to use it in the standard least squares approximations—now we can understand why squares and not other powers—because minimizing variance minimizes the entropy of a Gaussian distribution and thus the amount of information needed to characterize it.

Consider a Gaussian $B$ with $\langle B \rangle = 0$ and $\langle B^2 \rangle = \sigma^2$. If we have one bit to represent it, apparently, the only information we can convey is the sign of $B$. The simplest approach is to encode positive/negative regions by numbers $\pm A$. To minimize the squared error, we need to choose $A = \pm \langle |B| \rangle = \pm \sigma \sqrt{2/\pi}$, which corresponds to

$$\mathcal{D}(1) = 2(2\pi)^{-1/2} \int_0^\infty \left( B - \sigma \sqrt{2/\pi} \right)^2 \exp[-B^2/2\sigma^2] \frac{dB}{\sigma}$$

$$= \sigma^2 (1 - 2/\pi). \tag{3.11}$$

Let us now turn the tables and ask what minimal rate $R$ is sufficient to provide for distortion not exceeding $\mathcal{D}$. This is called the *rate-distortion function*, $R(\mathcal{D})$. We know that the rate is the mutual information $I(A, B)$, but now we are looking not for its maximum (as in channel capacity) but for the minimum over all the encodings defined by the conditional probabilities $P(B|A)$, such that the distortion does not exceed $\mathcal{D}$. Since $I(A, B) = S(B) - S(B|A)$, then minima of $I(A, B)$ are maxima of $S(B|A)$. It is helpful to think of distortion as produced by the added noise $\xi$ with the variance $\mathcal{D}$. For a fixed variance, maximal entropy $S(B|A)$ corresponds to the Gaussian distribution so that we have an (imaginary) Gaussian channel with the variance $\langle (B - A)^2 \rangle = \mathcal{D}$. Together with the Gaussian input having $\langle B^2 \rangle = \sigma^2$, they provide for the minimal rate given by (2.22):

$$R(\mathcal{D}) = I(A, B) = S(B) - S(B|A) = S(B) - S(B - A|A)$$

$$\geq S(B) - S(B - A) = \frac{1}{2} \log_2 \left( 2\pi e \sigma^2 \right) - \frac{1}{2} \log_2 \left( 2\pi e \mathcal{D} \right) = \frac{1}{2} \log_2 \frac{\sigma^2}{\mathcal{D}}.$$

$$\tag{3.12}$$

This goes to infinity for $\mathcal{D} \to 0$ and turns into zero for $\mathcal{D} = \sigma^2$. For $\mathcal{D} \geq \sigma^2$, we can take $A = 0$ with probability 1, making the mutual information zero—an absolute minimum! Note that stochastic encoding provides $D(1) = \sigma^2/4$ in (3.12), which is less than (3.11).

Often we need to represent by $R$ bits $m$ independent Gaussian signals with different variances $\sigma_i$, $i = 1, \ldots, m$—for instance, signals from different spectral intervals. How to divide these bits between signals to minimize the total distortion? We look for the distortions $\mathcal{D}_i$ and minimize $\sum_i \mathcal{D}_i = \mathcal{D}$ under the condition that $\sum_i R(\mathcal{D}_i) = R$. Differentiating $\sum_i [\mathcal{D}_i + \lambda \log_2 \sigma_i^2 / \mathcal{D}_i]$ with respect to $\mathcal{D}_i$, we find out that the total distortion is minimal when $\mathcal{D}_i$ are all equal, $\mathcal{D}_i = \mathcal{D}/m$, as long as this constant is less than all $\sigma_i$. Taking smaller $R$, we increase $\mathcal{D}$ and reach the moment when $\mathcal{D}/m$ exceeds some $\sigma_j$—then we need to take $R_j = 0$, that is, allocate zero bites to this component. Alternatively, if we manage to decrease enough of the variance of some component, it is not treated as fluctuating and does not deserve to be represented (except one bit for its mean if it is nonzero)—such is the logic of rate-distortion theory.

One can show that the rate-distortion function $R(\mathcal{D})$ is monotonous and convex for all systems. When the distortion is not a quadratic function, the conditional probability of encoding $P(A|B))$ is not Gaussian. In solving practical problems, it must be found by solving the variational problem, where one finds a normalized $P(A|B))$, which minimizes the mutual information under the condition of a given mean distortion. For that, one minimizes the functional

$$F = I + \beta \mathcal{D} = \sum_{ij} P(A_i|B_j) P(B_j) \left[ \ln \frac{P(A_i|B_j)}{P(A_i)} + \beta d(A_i, B_j) \right].$$

After variation with respect to $P(A_i|B_j)$ and enforcing normalization, we obtain

$$P(A_i|B_j) = \frac{P(A_i)}{Z(B_j, \beta)} e^{-\beta d(A_i, B_j)}, \tag{3.13}$$

where the partition function $Z(B_j, \beta) = \sum_i P(A_j) e^{-\beta d(A_i, B_j)}$ is the normalization factor. Recall that what is given is $P(B)$, not $P(A)$. The latter must be expressed via the same conditional probability:

$$P(A_i) = \sum_j P(A_i|B_j) P(B_j). \tag{3.14}$$

The system of linear equations (3.13, 3.14) is usually solved by iterations.

An immediate physical analogy is that (3.13) is a Gibbs distribution with the "energy" equal to the distortion function. Maximizing entropy for a given energy (Gibbs) is equivalent to minimizing mutual information for a given distortion function. As usual, what is given is in the exponent. The choice of the value of the inverse temperature $\beta$ reflects our priorities: at small $\beta$, the conditional probability is close to the unconditional one; that is, we minimize information without much regard to the distortion. On the contrary, large $\beta$ requires our conditional probability to be sharply peaked at the minima of the distortion function.

Similar but more sophisticated optimization procedures are applied, in particular, in image processing. Images contain an enormous amount of information. The rate at which visual data are collected by the photoreceptor mosaic of animals and humans is known to exceed $10^6$ bits/sec. On the other hand, studies on the speed of visual perception and reading speeds give numbers around 40–50 bits/sec for the perceptual capacity of the visual pathway in humans. The brain then has to perform huge data compressions. This is possible because visual information is highly redundant due to strong correlations between pixels.

The measured quantity $A$ thus contains too much data of low information value. We wish to compress $A$ to $C$ while keeping as much information as possible about $B$. Understanding the given signal $A$ requires more than just predicting/inferring $B$; it also requires specifying which features of the set of possible signals $\{A\}$ play a role in the prediction. Here meaning seeps back into information theory. Indeed, information is not knowledge (and knowledge is not wisdom). Not surprisingly, the main tool in automated and AI-assisted pattern recognition in images and other data is mutual information. We formalize this problem as one of finding a short code for $\{A\}$ that preserves the maximum information about the set $\{B\}$. That is, we squeeze the information that $A$ provides about $B$ through a "bottleneck" formed by a limited set of codewords $\{C\}$. This is reached via the method called the information bottleneck (Tishby at al. 2000), targeted at characterizing the trade-off between information preservation (accuracy of relevant predictions) and compression. Here one looks for the minimum of the functional

$$I(C, A) - \beta I(C, B). \tag{3.15}$$

The coding $A \rightarrow C$ is also generally stochastic, characterized by $P(C|A)$. The quality of the coding is determined by the rate, that is, by the average number of bits per message needed to specify an element in the codebook without confusion. This number per element $A$ of the source space $\{A\}$ is bounded from below by the mutual information $I(C, A)$, which we thus want to minimize. Effective coding utilizes the fact that mutual information is usually subextensive, in contrast to entropy, which is extensive. In section 2.4, we *maximized* $I(A, B)$ over all choices of the source space $\{B\}$ to find the channel capacity (upper bound for the error-free rate), while now we *minimize* $I(C, A)$ over all choices of coding. To put it differently: Before, we wanted to maximize the information transmitted, and now we want to minimize the information processed. This minimization, however, must be restricted by the need to retain in $C$ the relevant information about $B$, which we denote as $I(C, B)$.

Having chosen what properties of $B$ we wish to stay correlated with the encoded signal $C$, we add the mutual information $I(C, B)$ with the Lagrange multiplier to the functional (3.15). The sign of $\beta$ (inverse temperature) is positive to have minimal coding $I(A, B)$ preserving maximal information $I(C, B)$; that is, $I(C, B)$ is treated similarly to the channel capacity. The single parameter $\beta$ again represents the trade-off between the complexity of the representation, measured by $I(C, A)$, and the accuracy of this representation, measured by $I(C, B)$. At $\beta = 0$, our quantization is the most sketchy possible—everything in $A$ is assigned to a single codeword in $C$ and $I(C, A) = 0$. As $\beta$ grows, we are pushed toward detailed quantization. By varying $\beta$, one can explore the trade-off between preservation of the meaningful information and compression at various resolutions. Comparing with the rate-distortion theory functional (3.13), we recognize that we are looking for the conditional probability of the mapping $P(C|A)$; that is, we explicitly want to treat some pixels $A_i$ as more relevant than others.

However, the constraint on the meaningful information is now nonlinear in $P(C|A)$, so this is a much harder variational problem. Indeed, (3.15) can be written as follows:

$$I(C, A) - \beta I(C, B) = \sum_{ij} P(C_j|A_i)P(A_i) \ln \frac{P(C_j|A_i)}{P(C_j)}$$

$$- \beta \sum_{jk} P(B_k|C_j)P(C_j) \left\{ \ln \frac{P(B_k|C_j)}{P(B_k)} \right\}. \quad (3.16)$$

The conditional probabilities of $A, B$ under a given $C$ are related by the Bayes rule

$$P(B_k|C_j) = \frac{1}{P(C_j)} \sum_i P(A_i)P(B_k|A_i)P(C_j|A_i), \tag{3.17}$$

where the conditional probability of the measurements, $P(B_k|A_i)$, is presumed to be known. The variation of (3.16) with respect to the encoding conditional probability, $P(C_j|A_i)$, now gives the equation (rather than an explicit expression):

$$P(C_j|A_i) = \frac{P(C_j)}{Z(A_i, \beta)} \exp\left[ -\beta \sum_k P(B_k|A_i) \log \frac{P(B_k|A_i)}{P(B_k|C_j)} \right]$$

$$= \frac{P(C_j)}{Z(A_i, \beta)} \exp\left\{ -\beta D[P(B|A)||P(B|C)] \right\}. \tag{3.18}$$

We see that the relative entropy $D$ between the two conditional probability distributions emerges as the effective distortion measure $\mathcal{D}$. Here $P(B|A)$ is the true (data-given) distribution and $P(B|C)$ is our compressed encoded version. The system of equations (3.17, 3.18) is also solved by iterations. For example, one minimizes $I(A, C) + \beta D[P(B|A)||P(B|C)]$ in alternating iterations first over $P(C|A)$, then over $P(C)$, then over $P(B|C)$, then repeating the cycle. Doing the compression procedure many times, $A \to C_1 \to C_2 \ldots$ is used in multilayered deep learning algorithms. Statistical physics helps in identifying phase transitions (with respect to $\beta$) and suggests the relation between processing from layer to layer and the renormalization group: features along the layers become more statistically decoupled as the layers get closer to the fixed point.

Practical problems of iterations and machine learning are closely related to fundamental problems in understanding and describing biological evolution. Here an important task is to identify classes of functions and mechanisms that are provably evolvable—they can logically evolve into existence over realistic time periods and within realistic populations, without any need for combinatorially unlikely events to occur. Quantitative theories of evolution in particular aim to quantify the complexity of the mechanisms that evolved, which is done using information theory.

***Exercise 3.2:*** Rate-distortion function of a binary source.

Consider a binary source, which generates $B = 1$ with probability $p < 1/2$ and $B = 0$ with probability $1 - p$. Define the distortion function $d(A, B) = \delta_{AB} - 1$, that is, zero when $A = B$ and unity otherwise (the so-called Hamming function). Find the rate-distortion function $R(D)$.

## 3.6  Information Is Money

This section is for those brave souls who decided to leave science or engineering for gambling. If you have read till this point, you must be well prepared for that.

Let us start with the simplest game: you can bet on a coin, doubling your bet if you are right or losing it if you are wrong. Surely, an intelligent person would not bet money hard saved during graduate studies on a totally random process with zero gain. You bet only when you have *information* that one side has the probability $p > 1/2$. To have an average growth, you want to play the game many times. Shall we look then for the maximal average return? The maximal mean arithmetic growth factor is $(2p)^N$ and corresponds to betting all your money every time on the more probable side. Such a mean, however, comes from a single all-win realization; the probability of that winning streak goes to zero with growing $N$ as $p^N$. To avoid losing it all with probability fast approaching unity, you bet only a fraction $f$ of your money on the more probable $p$ side. What to do with the remaining money—keep it as insurance or bet on a less probable side? The first option just diminishes the effective amount of money that works. Moreover, the other side also wins sometimes, so we put $1 - f$ on the side with $1 - p$ chance. If after $N$ such bets the $p$ side wins $n$ times, then your money is multiplied by the factor $(2f)^n[2(1-f)]^{N-n} = 2^{N\Lambda}$, where the rate is

$$\Lambda(f) = 1 + \frac{n}{N} \log_2 f + \left(1 - \frac{n}{N}\right) \log_2(1 - f). \qquad (3.19)$$

As $N \to \infty$, we approach the mean geometric rate, which is $\lambda = 1 + p \log f + (1 - p) \log(1 - f)$. Note the similarity with the Lyapunov exponents that are introduced in section 5.3—we consider the logarithm of the exponentially growing factor since we know $\lim_{N \to \infty}(n/N) = p$ (it is called a self-averaging quantity because it is again a sum of random numbers). Differentiating $\Lambda(f)$ with respect to $f$, you find that the maximal growth rate corresponds to $f = p$ (proportional gambling) and equals

$$\lambda(p) = 1 + p \log_2 p + (1 - p) \log_2(1 - p) = S(u) - S(p), \qquad (3.20)$$

where we denote the entropy of the uniform distribution $S(u) = 1$ bit. We thus see that the maximal rate of money growth equals the entropy decrease, that is, the information you have (Kelly 1950). What is beautiful here is that the proof of optimality is constructive and gives us the best betting strategy. An important lesson is that we maximize not the mean return but its mean logarithm, which is a geometric mean. Since it is a self-averaging quantity, the probability to grow with this rate approaches unity as $N \to \infty$. Note, however, that the geometric mean is less than the arithmetic mean because the logarithm is a convex function. Therefore, we may have a situation when the arithmetic growth factor is larger than unity while the geometric mean is smaller than unity. Don't play such games, since no matter the strategy, the probability of losing it all tends to unity as $N \to \infty$, even though the mean returns grow unbounded.

It is straightforward to generalize (3.20) for gambling on horse races or investing, where many outcomes have different probabilities $p_i$ and payoffs $g_i$. To maximize $\sum p_i \log(f_i g_i)$, we look for the maximum of $\sum p_i \log f_i$. Since $\sum f_i = 1$, we can treat it as a distribution. The relative entropy $\sum p_i \log(p_i/f_i)$ is nonnegative so that $\sum p_i \log f_i$ reaches its maximum when all $f_i = p_i$ independent of $g_i$; that is, our distribution coincides with the true distribution, which is proportional gambling. The rate is then

$$\lambda(p, g) = \sum_i p_i \ln(p_i g_i). \qquad (3.21)$$

Here you have a formidable opponent—the track operator, who actually sets the payoffs. Knowing the probabilities, a perfect (idealistic) operator would set the payoffs, $g_i = 1/p_i$, to make the game fair and your rate zero. Nobody's perfect, so it is more likely that a realistic operator has the business sense to make the racecourse profitable by setting the payoffs a bit lower. That will make your $\lambda$ negative. For example, the European roulette wheel has 18 red and 18 black pockets and a single green, so that even the highest-odds bets, on red or black, have a slightly less than half chance of success.

Your only hope then is that your information is better. If the operator assumes that the probabilities are $q_i$ and sets payoffs as $g_i = 1/Zq_i$ with $Z > 1$, then

$$\lambda(p, q) = -\ln Z + \sum_i p_i \ln(p_i/q_i) = -\ln Z + D(p|q). \qquad (3.22)$$

That is, if you know the true distribution but the operator uses the approximate one, the relative entropy $D(p|q)$ determines the rate with which your winnings can grow. Since you aren't perfect either, then it is likely that you use the distribution $q'$, which is not the true one. In this case, you still have a chance if your distribution is closer to the true one: $\lambda(p, q, q') = -\ln Z + D(p|q) - D(p|q')$. Recall that the entropy determines the optimal rate of coding. Using an incorrect distribution incurs the cost of nonoptimal coding. Amazingly, (3.22) tells us that if you can encode the data describing the sequence of track winners so that it is shorter than the operator's, you get paid in proportion to that shortening.

That beautiful theory, of course, has nothing to do with how people bet and bookmakers operate. In reality, people bet according to their whims rather than by playing a long game, while bookmakers set the rewards according to the statistics of betting rather than horse winnings. Average gambler losses and bookmaker income are independent of the outcome of racing, which is thus a pure sport (most of the time).

The theory, however, has found numerous applications in engineering and biology. It turns out that bacteria follow the proportional gambling strategy without ever taking this or another course on information theory. Like in coin flipping, bacteria face the choice, for instance, between growing fast but being vulnerable to antibiotics or growing slow but being resistant. They use proportional gambling to allocate respective fractions of populations to different choices. There could be several lifestyle choices, analogous to horse racing, called phenotype switching in this case. The same strategy is used by many plants, where a fraction of the seeds do not germinate in the same year they were dispersed; the fraction increases with environmental variability.

More generally, the environment can be characterized by a set of parameters $A$, while the internal state of a gambler, plant, or bacteria can be characterized by another set of parameters $B$. In the proportional gambling setting, $A$ is the vector of probabilities $\{p_i\}$ and $B$ is the vector of fractions $\{f_i\}$. In another setting, $A$ could include the concentration of a nutrient and $B$ the amount of an enzyme needed to metabolize the nutrient. The logarithmic growth rate is then the function of these two parameters, $r(A, B)$, and the mean rate is as follows:

$$\lambda = \int dA \, dB \, P(A, B) r(A, B) = \int dA \, P(A) \int dB \, P(B|A) r(A, B). \quad (3.23)$$

To maximize growth, bacteria, plants, and gamblers need to coordinate their internal state with that of the environment. That coordination is governed by the conditional probability $P(B|A)$, which determines the mutual information between the external world and the internal state:

$$I(A, B) = \int dA \, P(A) \int dB P(B|A) \log_2 \frac{P(B|A)}{P(B)}. \quad (3.24)$$

But acquiring that information has its own cost, $aI$. One then looks for a trade-off between maximizing growth and minimizing information cost. Therefore, we look for the maximum of the functional $F = \lambda - aI$, which gives similarly to (3.13)

$$P(B|A) = \frac{P(B)}{Z(A, \beta)} e^{\beta r(A,B)}, \qquad (3.25)$$

where $\beta = a^{-1} \ln 2$ and the partition function $Z(A, \beta) = \int dB P(B) e^{\beta r(A,B)}$ is the normalization factor. We now recognize the rate-distortion theory from the previous section; the only difference is that the energy now is minus the growth rate. The choice of $\beta$ reflects the relative costs of the information and the metabolism. If information is hard to get, one chooses small $\beta$, which makes $P(B|A)$ weakly dependent on $r(A, B)$ and close to unconditional probability $P(B)$. If information is cheaper, (3.25) tells us that we need to peak our conditional probability around the maxima of the growth rate. All the possible states in the plane $r, I$ are below some monotonic convex curve, much like in the energy-entropy plane in section 1.1. One can reach the optimal (Gibbs) state on the boundary either by increasing the growth rate at a fixed information or by decreasing the information at a fixed growth rate.

Section A.6 describes a remarkable class of strategies to find an optimal balance between exploration for new information on the environment and exploitation of the existing information to maximize growth.

Financial activity is not completely reducible to gambling and its essence is understood much less. When you earn enough money (or no money at all), it may be a good time to start thinking about the nature of money itself. Money appeared first as a measure of value and acquired a probabilistic aspect with the development of credit. These days, when most of it is in bits, it is clear that money is less matter (coins, banknotes) and more information. The total amount of money grows on average but could experience sudden drops when a crisis arrives. Yet in payments money behaves as energy and matter, satisfying the conservation law. It seems that we need a new concept for describing money, which has properties of both entropy and energy. Free energy combines energy and entropy additively, describing, in particular, how an entropy increase (loss of information) diminishes the amount of work one can do. Similarly, free energy can describe a decrease in purchasing power due to information loss. However, to describe money as a universal medium of exchange, we probably need a more sophisticated notion. Since money is essentially a social construct, the degree of universality varies. For example, cold hard cash is guaranteed by governments, but credit card payments are guaranteed by private banks, so these two kinds of money are not identical. Add to this nonbank money like cryptocurrencies, and we see that the value of money depends essentially on how many people agree to use it. It is a challenge to devise a conceptual framework able to handle both the material and ephemeral sides of money, but it seems that information theory is the right place to start.

**Exercise 3.3:**  Bookmaker's sure bet.

In a series of two-horse races, the first horse wins three times more often than the second one. Yet public sentiment is such that it bets on the first horse only twice as many times. A bookmaker has two choices to set the rewards: i) according to race probabilities, pay respectively $4/3Z$ and $4/Z$ times the amount of the bet on the first/second horse, ii) according to public preferences, pay respectively $3/2Z$ and $3/Z$ times the amount of the bet on the first/second horse. Here $Z > 1$ to guarantee a profit. Which strategy is preferable?

# 4

# New Second Law of Thermodynamics

So far, we have quantified uncertainty mostly by combinatorics. Classifying and keeping count are among the most difficult mental processes (possibly because they require impartiality and memory). It is best to hire somebody else to do the job. That tireless somebody, who never stops, is a random walker. In this chapter, we exploit the walker and explore random walks in different environments. We first use a random walk on a graph to describe Google's PageRank algorithm, designed to quantify not the amount of information but its perceived importance. We then consider a random walk on a lattice biased by an externally imposed time-dependent drift. That leads us to fluctuation-dissipation relations and the modern generalizations of the second law of thermodynamics. This is important both for the fundamentals of science and for numerous modern applications related to fluctuations in nanoparticles, macromolecules, stock market prices, etc.

## 4.1   Stochastic Web Surfing and Google's PageRank

When it was proclaimed that the Library contained all books, the first impression was one of extravagant happiness . . . followed by an excessive depression. The certitude that some precious books were inaccessible seemed almost intolerable.

—JORGE LUIS BORGES, "THE LIBRARY OF BABEL"

To know which are the most precious books in the library, we need an objective and quantitative measure of information importance. For efficient information retrieval from the web library, web pages need to be ranked by

81

their importance to order search results. By this time, it should come as no surprise for the reader that such ranks can be found by a statistical approach: performing a random walk on the web.

For an internet with $n$ pages, we organize their ranks into a vector $\mathbf{p} = \{p_1, \ldots, p_n\}$, which we normalize: $\sum_{i=1}^{n} p_i = 1$. The idea is to *equate the rank $p_i$ with the probability to arrive at this page* by randomly clicking on links. A reasonable way to measure the probability of arriving at a page is to count the number of links that refer to it. Not all links are equal, though—those from a more probable page must bring more probability. On the other hand, a link from a page with many outgoing links must bring to each link less probability. One then comes to the two rules: i) every page relays its rank to the pages it links to, dividing it equally between them, and ii) the rank of a page is the sum of all ranks obtained by links. According to these rules, $p_i = \sum_j p_j/n_j$, where $n_j$ is the number of outgoing links on page $j$, which links to page $i$. In other words, we are looking for the eigenvector of the hyperlink matrix, $\mathbf{p}\hat{A} = \mathbf{p}$, where the matrix elements $a_{ij} = 1/n_j$ if $j$ links to $i$ and $a_{ij} = 0$ otherwise. Does a unique eigenvector with all non-negative entries and a unit eigenvalue always exist? If yes, how to find it?

The iterative algorithm to find the rank eigenvector $p_i$ is called PageRank (Brin and Page 1998).[1] It starts by ascribing equal probability to all pages, $p_i(0) = 1/n$, and generates the new probability distribution by applying the above rules of the rank relay:

$$\mathbf{p}(t+1) = \mathbf{p}(t)\hat{A}. \qquad (4.1)$$

We recognize that this stochastic process is a Markov chain, mentioned in sections 2.3 and 3.6, which means that the future is determined by the present state, but not by the past. We thus interpret $\hat{A}$ as the matrix of transition probabilities between pages for our random surfer. In later modifications, rather than fill the elements of $\hat{A}$ uniformly as $1/n_j$, one uses information about actual frequencies of linking that can be obtained from access logs. Could our self-referential rules lead to a vicious circle or the iterations converge at $t \to \infty$? It is clear that if the largest eigenvalue $\lambda_1$ of $\hat{A}$ is larger than unity, then the iterations would diverge; if $\lambda_1 < 1$, then the iterations would converge to zero. Both contradict normalization, $\sum p_i = 1$. We need the largest eigenvalue to be unity and correspond to a single eigenvector so that the iterations converge.

---

1. "Page" relates both to web pages and to Larry Page, who with Sergei Brin invented the algorithm and created Google.

How fast it converges is then determined by the second largest eigenvalue $\lambda_2$ (which must be less than unity).

A moment's reflection is enough to identify the problem: some pages do not link to any other page, which corresponds to rows of off-diagonal zeros in $\hat{A}$. Such pages accumulate the score without sharing it. Another problem is caused by loops. The figure presents a simple example illustrating both problems:

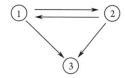

If all transition probabilities are nonzero, the probability vector with time tends to $(0, 0, 1)$, that is, the surfer is stuck at page 3. When the probabilities $a_{13}, a_{23}$ are very small, the surfer tends to be caught for long times in the loop $1 \longleftrightarrow 2$.

To release our random surfer from being stuck at a sink or caught in a loop, the original PageRank algorithm allowed it to jump randomly to any other page with equal probability. To be fair with pages that are not sinks, these random teleportations are added to all nodes in the web: the surfer either clicks on a link on the current page with probability $d$ or opens up a random page with probability $1 - d$. To quote the original: "We assume there is a 'random surfer' who is given a web page at random and keeps clicking on links, never hitting 'back' but eventually gets bored and starts on another random page. The probability that the random surfer visits a page is its PageRank. The damping factor is the probability that the 'random surfer' will get bored and request another random page." This is equivalent to replacing $\hat{A}$ by $\hat{G} = d\hat{A} + (1 - d)\hat{E}$. Here the teleportation matrix $\hat{E}$ has all entries $1/n$, that is, $\hat{E} = \mathbf{e}\mathbf{e}^T/n$, where $\mathbf{e}$ is the column vector with $e_i = 1$ for $i = 1, \ldots, n$. After that, all matrix entries $g_{ij}$ are strictly positive and the graph is fully connected.

It is important that our matrix now has positive elements in every column whose sum is unity. Such matrices are called stochastic since every column can be thought of as a probability distribution. Every stochastic matrix has unity as the largest eigenvalue. Indeed, since $\sum_j g_{ij} = 1$, then $\mathbf{e}$ is an eigenvector of the transposed matrix: $\hat{G}^T \mathbf{e} = \mathbf{e}$. Therefore, 1 is an eigenvalue for $\hat{G}^T$, and also for $\hat{G}$, which has the same eigenvalues. We can now use convexity to prove that this is the largest eigenvalue. For any vector $\mathbf{p}$, every element of $\mathbf{p}\hat{G}$ is a convex combination of the elements, $\sum_j p_j g_{ij}$, which cannot exceed the largest element of $\mathbf{p}$ since $\sum_j g_{ij} = 1$. For an eigenvector with an eigenvalue exceeding unity, at least one element of $\mathbf{p}\hat{G}$ must exceed the largest element of $\mathbf{p}$; therefore, such an eigenvector cannot exist. This is a particular case of the following theorem: The eigenvalue with the largest absolute value of a positive square matrix is positive and belongs to a positive eigenvector, where all of the vector's elements are positive. All other eigenvectors are smaller in absolute value (Markov 1906, Perron 1907).

The great achievement of the PageRank algorithm is that it replaces the process (4.1) by

$$\mathbf{p}(t+1) = \mathbf{p}(t)\hat{G}. \tag{4.2}$$

That iterative process cannot be caught into a loop and converges, which follows from the fact that $G_{ii} \neq 0$ for all $i$; that is, there is always a probability of staying on the page, breaking any loop. The eigenvalues of $\hat{G}$ are $1, d\lambda_2 \ldots d\lambda_n$, where $\{\lambda_i\}$ are eigenvalues of $\hat{A}$ (prove it), so the choice of $d$ affects convergence, the smaller the faster. On the other hand, it is somewhat artificial to use teleportation to an arbitrary page, so larger values of $d$ give more weight to the true link structure of the web. As in other optimization problems we have encountered in this book, a workable compromise is needed. The standard Google choice, $d = 0.85$, comes from estimating how often an average surfer uses bookmarks. As a result, the process usually converges after about 50 iterations.

One can design a personalized ranking by replacing the teleportation matrix by $\hat{E} = \mathbf{e}\mathbf{v}^T$, where the probability vector $\mathbf{v}$ has all nonzero entries and allows for personalization; that is, it can be chosen according to the individual user's history of searches and visits. That means that it is possible in principle to have our personal rankings of the web pages and make searches customized.

As mentioned, the sequence of the probability vectors defined by the relations of the type (4.1, 4.2) is a Markov chain. In particular, the three random quantities $X \to Y \to Z$ constitute a Markov triplet if $Y$ is completely determined by $X, Z$, while $X, Z$ are independent, conditional on $Y$; that is, $I(X, Z|Y) = 0$. Such chains have an extremely wide domain of applications.

***Exercise 4.1:*** PageRank of the two-page internet.

Consider the simplest version of the internet, which has two pages: page 1 has one link to page 2, which has no links. Rank these pages according to the PageRank algorithm with arbitrary $d < 1$ (the probability of following a link).

**Exercise 4.2:** Eigenvalues of the Google matrix.

Assume that the matrix $\hat{A}$ with the spectrum $(1, \lambda_2, \ldots, \lambda_n)$ is stochastic, that is, $\sum_j a_{ij} = 1$ for every $i$. Prove that the spectrum of the Google matrix $\hat{G} = d\hat{A} + (1 - d)\mathbf{ev}^T$ is $(1, d\lambda_2, \ldots, d\lambda_n)$, where $\mathbf{v}$ is an arbitrary probability vector, that is, $\sum_i v_i = 1$.

**Exercise 4.3:** Solus Rex. A king randomly moves to any of the adjacent squares with equal probability on an otherwise empty $3 \times 3$ chessboard.

(a) How much information brings a message specifying his position?

(b) If we wish to encode the whole game (the random walk of the king), we need to know how the number of typical sequences $N(n)$ grows asymptotically with the number $n$ of the moves: $\lim_{n \to \infty} N(n) = 2^{nS}$. Find $S$, which is called the information rate of the source. Is it the same as the entropy that determined the answer to the previous question?

## 4.2   Random Walk and Diffusion

Let us now consider a particular yet fundamental stochastic process of a random walk on a lattice, where the transition probability is nonzero only for neighboring sites. Our walker can hop randomly to any of the $2d$ neighboring sites in the $d$-dimensional cubic lattice, starting from the origin at $t = 0$. We denote $a$ as the lattice spacing, $\tau$ as the time between hops, and $\mathbf{e}_i$ as the orthogonal lattice vectors that satisfy $\mathbf{e}_i \cdot \mathbf{e}_j = a^2 \delta_{ij}$. The probability of being in a given site $\mathbf{x}$ evolves according to the equation

$$P(\mathbf{x}, t + \tau) = \frac{1}{2d} \sum_{i=1}^{d} \left[ P(\mathbf{x} + \mathbf{e}_i, t) + P(\mathbf{x} - \mathbf{e}_i, t) \right]. \tag{4.3}$$

That can be rewritten in the form convenient for taking the continuous limit:

$$\frac{P(\mathbf{x}, t + \tau) - P(\mathbf{x}, t)}{\tau} = \frac{a^2}{2d\tau} \sum_{i=1}^{d} \frac{P(\mathbf{x} + \mathbf{e}_i, t) + P(\mathbf{x} - \mathbf{e}_i, t) - 2P(\mathbf{x}, t)}{a^2}. \tag{4.4}$$

This is a finite difference approximation to the diffusion equation, which appears when we take the continuous limit $a \to 0$, $\tau \to 0$, keeping finite the ratio $\kappa = a^2/2d\tau$: $(\partial_t - \kappa \Delta)P(\mathbf{x}, t) = 0$. The space density $\rho(x, t) = P(\mathbf{x}, t)a^{-d}$

satisfies the same equation:

$$\partial_t \rho = \kappa \Delta \rho. \tag{4.5}$$

The solution with the initial condition $\rho(\mathbf{x}, 0) = \delta(\mathbf{x})$ is the Gaussian distribution:

$$\rho(\mathbf{x}, t) = (4\pi \kappa t)^{-d/2} \exp\left(-x^2/4\kappa t\right). \tag{4.6}$$

Along with (4.3) and (4.4), the diffusion equation conserves the total probability, $\int \rho(\mathbf{x}, t)\, d\mathbf{x}$, because it has the form of a continuity equation, $\partial_t \rho(\mathbf{x}, t) = -\mathrm{div}\, \mathbf{j}$ with the current $\mathbf{j} = -\kappa \nabla \rho$.[2] Note that (4.5) and (4.6) are isotropic and translation invariant, while the discrete version respects only cubic symmetries. The phase volume increases as $t^{d/2}$, and the entropy grows with time logarithmically.

Another way to describe a random walk is to treat $\mathbf{e}_i$ as a random variable with $\langle \mathbf{e}_i \rangle = 0$ and $\langle \mathbf{e}_i \mathbf{e}_j \rangle = a^2 \delta_{ij}$, so that $\mathbf{x} = \sum_{i=1}^{t/\tau} \mathbf{e}_i$. The probability of the sum is (4.6), that is, the product of Gaussian distributions of the components, with the variance growing linearly with $t$.

A path of a random walker behaves more like a surface than a line. The two-dimensionality of the random walk is a reflection of the square-root diffusion law: $\langle x \rangle \propto \sqrt{t}$. The dimensionality of a set defines the relation between its size $x$ and the number $N \propto x^d$ of standard-size elements needed to cover it. For a random walk, the number of elements is on the order of the number of steps, $N \propto t \propto x^2$. One can also look at how the number of boxes $N(a)$ needed to cover a geometric object grows as the box size $a$ decreases; see the definition of the box-counting dimension (5.20) in section 5.4. For a line $N \propto 1/a$, generally $N \propto a^{-d}$. As we discussed above, diffusion requires the time step to shrink with the lattice spacing according to $\tau \propto a^2$. The number of elements is the number of steps and grows for a given $t$ as $N(a) = t/\tau \propto a^{-2}$ so that $d = 2$. Surfaces generally intersect along curves in 3D; they meet at isolated points in 4D and do not meet at $d > 4$. That is reflected in special properties of the random walk in 2D (where it fills the surface) and 4D (where random walkers do not meet and hence do not interact). The mean time spent on a given site, $\sum_{t=0}^{\infty} P(\mathbf{x}, t) \to \int \rho(\mathbf{x}, t)\, dt \propto \int^{\infty} t^{-d/2} dt$, diverges for $d \leq 2$. In other words, the walker in 1D and 2D returns to any point an infinite number of times.

**Exercise 4.4:** Random walk on a circle.

Consider a one-dimensional random walk over a circle with $N$ sites as a Markov chain and write the one-step transformation of the probability

---

2. The continuity equation is derived in detail at the beginning of section 5.1.

distribution over the sites $i = 1, \ldots, N$. Find the transition probability matrix $\hat{A}$ and show that its eigenvectors are $e^{ijk}$ if $k_n = 2\pi n/N$ for $n = 0, 1, \ldots, N - 1$. Show that the only stationary distribution is the eigenvector with the highest eigenvalue and the rate of relaxation to it is determined by the second largest eigenvalue.

## 4.3 Detailed Balance

A significant generalization of equilibrium statistical physics can be achieved for systems with one or few degrees of freedom deviated arbitrarily far from equilibrium. That can be done under the assumption that the rest of the degrees of freedom are in equilibrium and can be represented by a thermostat generating thermal noise. This new approach also allows one to treat nonthermodynamic fluctuations, such as a negative entropy change.

We illustrate these developments using the simplest example of a one-dimensional random walk, to which we add a drift with velocity $v(x) = -\partial V(x)/\partial x$; that is, down the gradient of the potential $V(x)$. According to the continuity equation, the drift adds $\rho v$ to the current $j$ and $-\partial_x(\rho v)$ to $\partial \rho(x, t)/\partial t$. Combining this with (4.5), which describes diffusion from the random walk, we obtain the so-called Fokker-Planck equation for the probability $\rho(x, t)$ (see section A.10 for the detailed derivation):

$$\partial_t \rho = T\partial_x^2 \rho + \partial_x(\rho\partial_x V) = -\partial_x j. \qquad (4.7)$$

Here we denote diffusivity as $T$ for reasons that will be clear in a moment. Let us denote $\rho(x', t; x, 0)$ as the conditional probability to come from $x$ at 0 to $x'$ at $t$. Without the coordinate-dependent field $V(x)$, the transition probability is symmetric:

$$\rho(x', t; x, 0) = \rho(x, t; x', 0). \qquad (4.8)$$

That property is called the detailed balance.

How is it modified in an external field? If the potential $V$ is time-independent, then the stationary solution of (4.7) is the zero-current Gibbs state: requiring $j = -T\partial_x \rho - \rho\partial_x V = 0$, we obtain

$$\rho(x) = Z_0^{-1} \exp[-\beta V(x)], \quad Z_0 = \int \exp[-\beta V(x, 0)]\, dx. \qquad (4.9)$$

From the perspective of information theory, if the only condition we impose is that a particle at $x$ has the mean energy $V(x)$, then the probability distribution is an exponent of the energy. The Gibbs state (4.9) satisfies a modified detailed balance: the probability current is the (Gibbs) probability density at the starting point times the transition probability; forward and backward currents must be equal in equilibrium:

$$\rho(x', t; x, 0)e^{-V(x)/T} = \rho(x, t; x', 0)e^{-V(x')/T}. \tag{4.10}$$

We can exploit an analogy between quantum mechanics and statistical physics by introducing the Fokker-Planck operator,

$$H_{FP} = -\frac{\partial}{\partial x}\left(\frac{\partial V}{\partial x} + T\frac{\partial}{\partial x}\right).$$

This operator governs the evolution of the probability density the same way the Hamiltonian governs the evolution of the quantum wave function. The probability density is then treated as a $\psi$ function in the $x$ representation, $\rho(x, t) = \langle x|\psi(t)\rangle$, using notations from chapter 6. We then rewrite (4.7) as $d|\psi\rangle/dt = -\hat{H}_{FP}|\psi\rangle$, which has the formal solution $|\psi(t)\rangle = \exp(-tH_{FP})|\psi(0)\rangle$. The only difference with quantum mechanics is that their time is imaginary (of course, they think that our time is imaginary). In other words, the Schrodinger equation,

$$\imath\hbar\frac{d}{dt}|\psi\rangle = \left(-\frac{h^2}{2m}\Delta + V\right)|\psi\rangle,$$

corresponds to imaginary diffusivity. The transitional probability is given by the matrix element:

$$\rho(x', t'; x, t) = \langle x'|\exp[(t - t')H_{FP}]|x\rangle. \tag{4.11}$$

The quantum-mechanical notations allow us to translate the detailed balance from the property of transition probabilities to that of the evolution operator. The field-free symmetry (4.8) formally corresponds to the fact that the respective Fokker-Planck operator $\partial_x^2$ is Hermitian. The relation (4.10) can be written as follows:

$$\langle x'|e^{-tH_{FP}}e^{-V/T}|x\rangle = \langle x|e^{-tH_{FP}}e^{-V/T}|x'\rangle = \langle x'|e^{-V/T}e^{-tH_{FP}^\dagger}|x\rangle.$$

Since this must be true for any $x, x'$, then $e^{-tH_{FP}^\dagger} = e^{V/T}e^{-tH_{FP}}e^{-V/T}$ and

$$H_{FP}^\dagger \equiv \left(\frac{\partial V}{\partial x} - T\frac{\partial}{\partial x}\right)\frac{\partial}{\partial x} = e^{V/T}H_{FP}e^{-V/T}, \tag{4.12}$$

i.e., $e^{V/2T}H_{FP}e^{-V/2T}$ is Hermitian, which can be checked directly (more on the analogy between thermal and quantum fluctuations can be found in section A.12).

## 4.4  Fluctuation Relation and the New Second Law

Two hundred years after Carnot, one might expect the second law of thermo-dynamics to be a bronze monument, yet it is very much alive, growing, and changing shape.

To see it, let us allow the potential $V$ in $(4.7)$ to change in time; then the system goes away from equilibrium. Consider an ensemble of trajectories starting from the initial positions taken with the probabilities determined by the equilibrium Gibbs distribution corresponding to the initial potential: $\rho(x, 0) = Z_0^{-1} \exp[-\beta V(x, 0)]$, where $\beta = 1/T$. As time proceeds and the potential continuously changes, the system is never in equilibrium, so that $\rho(x, t)$ does not generally have a Gibbs form. Even though one can define a time-dependent Gibbs state, $Z_t^{-1} \exp[-\beta V(x, t)]$, with $Z_t = \int \exp[-\beta V(x, t)]dx$, one can directly check that it is not a solution of the Fokker-Planck equation $(4.7)$ because of the extra term: $\partial_t \rho = -\beta \rho \partial_t V$. The distribution needs some time to adjust to the potential changes and is generally dependent on the history of these. For example, if we suddenly broaden the potential well to the width $L$, it will take diffusion (with diffusivity $T$) a time of order $L^2/T$ to broaden the distribution. Can we find some quantity that accounts for this history and lets us generalize the detailed balance relation $(4.10)$ we had in equilibrium? Such a relation was found surprisingly recently despite its generality and relative technical simplicity of derivation.

To find the quantity that has a Gibbs form (i.e., that has its probability determined by the instantaneous potential), we need to find an equation that generalizes $(4.7)$ by having an extra term that will cancel the time derivative of the potential. This is achieved by considering, apart from a position $x$, another random quantity, defined as the potential energy change (or the external work done) along the particle trajectory during time $t$:

$$W_t = \int_0^t dt' \frac{\partial V(x(t'), t')}{\partial t'}. \qquad (4.13)$$

The time derivative is partial, i.e., taken only with respect to the second argument, so that the integral is not equal to the difference between the start and the finish, but is determined by the whole history. The work is a fluctuating and even sign-changing quantity depending on the trajectory $x(t')$, which itself depends on the initial point and random walk realization.

Let us now run our random walker many times, choosing different starting points $x(0)$ according to the Gibbs probability $\rho(x) = Z_0^{-1} \exp[-\beta V(x, 0)]$.

This gives us many trajectories having different endpoints $x(t)$ and accumulating different energy changes $W$ along the way. Now consider the joint probability $\rho(x, W, t)$ of reaching $x$ and acquiring energy change $W$. This two-dimensional probability distribution satisfies the generalized Fokker-Planck equation, which can be derived as follows: similar to the argument preceding (4.7), we note that the flow along $W$ in $x - W$ space proceeds with the velocity $dW/dt = \partial_t V$, so that the respective component of the current is $\rho \partial_t V$ and the equation takes the form

$$\partial_t \rho = \beta^{-1} \partial_x^2 \rho + \partial_x(\rho \partial_x V) - \partial_W \rho \partial_t V. \tag{4.14}$$

Since $W_0 = 0$, the initial condition for (4.14) is

$$\rho(x, W, 0) = Z_0^{-1} \exp[-\beta V(x, 0)]\delta(W). \tag{4.15}$$

While we cannot find $\rho(x, W, t)$ for arbitrary $V(t)$, we can multiply (4.14) by $\exp(-\beta W)$ and integrate over $dW$. Since $V(x, t)$ does not depend on $W$, we get the closed equation for $f(x, t) = \int dW \rho(x, W, t) \exp(-\beta W)$:

$$\partial_t f = \beta^{-1} \partial_x^2 f + \partial_x(f \partial_x V) - \beta f \partial_t V. \tag{4.16}$$

Now, *this* equation does have an exact time-dependent solution,

$$f(x, t) = Z_0^{-1} \exp[-\beta V(x, t)],$$

where the factor $Z_0^{-1}$ is chosen to satisfy the initial condition (4.15). Note that $f(x, t)$ is instantaneously defined by $V(x, t)$ without any history dependence, in contrast to $\rho(x, t)$. In other words, the distribution weighted by $\exp(-\beta W_t)$ looks like the Gibbs state, adjusted to the time-dependent potential at every moment of time. Even though the phase volume defines probability only in equilibrium, the work divided by temperature is an analog of the entropy change (production), and the exponent of it is an analog of the phase volume change. Let us stress that $f(x, t)$ is not a probability distribution. In particular, its integral over $x$ is not unity but the mean phase volume change, which remarkably is expressed via equilibrium partition functions at the ends (Jarzynski 1997):

$$\int f(x, t)dx = \int \rho(x, W, t)e^{-\beta W} dx dW = \langle e^{-\beta W} \rangle = \frac{Z_t}{Z_0} = \frac{\int e^{-\beta V(x,t)} dx}{\int e^{-\beta V(x,0)} dx}.$$

$$\tag{4.17}$$

Here the bracket means double averaging: over the initial distribution $\rho(x, 0)$ and over the different random walks during the time interval $(0, t)$. We can also obtain all weighted moments of $x$, like $\langle x^n \exp(-\beta W_t) \rangle$.[3] One can introduce the free energy $F_t = -T \ln Z_t$, so that $Z_t/Z_0 = \exp[\beta(F_0 - F_t)]$.

Let us reflect on where we have arrived following our random walker. We started from a Gibbs distribution but considered *arbitrary* temporal evolution of the potential. Therefore, our distribution was arbitrarily far from equilibrium during the evolution. Despite that, we expressed the mean exponent of the work done via the partition functions of the equilibrium distributions, corresponding to the potential at the beginning and at the end. Even though the system is not in equilibrium at the end, the use of the Gibbs distribution is not that surprising, because the further relaxation to equilibrium at the end value of the potential is not accompanied by doing any work $W$. What is surprising is that there is no dependence on the intermediate times. One can also look at it from the opposite perspective: no less remarkable is that one can determine a truly equilibrium property, the free energy difference, from nonequilibrium measurements (which could be arbitrarily fast rather than adiabatically slow, as we used to do in traditional thermodynamics).

The total heat release is the work minus the free energy change: $Q = W - F_t + F_0$. Divided by the temperature, this is minus the entropy change during the evolution. That allows us to rewrite (4.17) as the following identity:

$$\langle e^{-Q/T} \rangle = \langle e^{-\Delta S} \rangle = 1, \tag{4.18}$$

which is a generalization of the second law of thermodynamics. Note that the entropy change $\Delta S$ is treated here as a fluctuating quantity, which could have either sign. Using the Jensen inequality $\langle e^A \rangle \geq e^{\langle A \rangle}$, one can obtain the usual second law of thermodynamics for the positivity of the mean entropy change:

$$\langle \Delta S \rangle \geq 0.$$

When information processing is involved, it must be treated on an equal footing, which allows one to decrease the work and the dissipation below the free energy difference (Sagawa and Uedo, 2012; Sagawa 2012):

$$\langle e^{-\beta Q - I} \rangle = \langle e^{-\Delta S} \rangle = 1. \tag{4.19}$$

We considered such a case in section 3.2, where we used $\langle Q \rangle \geq -IT = -T\Delta S$. The exponential equality (4.19) is a generalization of this inequality and (3.7).

---

3. I thank R. Chetrite for this derivation.

So the modern form of the second law of thermodynamics is equality rather than inequality. The latter is just a partial consequence of the former. Compare it with the reformulation of the second law in section 5.2 as a conservation law rather than a law of increase.

And yet (4.19) is not the most general form. The further generalization is achieved by relating the entropy production to irreversibility, stating that the probability of having a change $-\Delta S$ in a time-reversed process (marked by a dagger) is as follows (Crooks 1999):

$$P^\dagger(-\Delta S) = P(\Delta S)e^{-\Delta S}. \qquad (4.20)$$

Integrating (4.20), one obtains (4.19). That remarkable relation also allows one to express the mean entropy production via the relative entropy (2.29) between probabilities of the forward and backward evolution:

$$\langle \Delta S \rangle = \left\langle \ln[P(\Delta S)/P^\dagger(-\Delta S)] \right\rangle. \qquad (4.21)$$

The positivity of the mean entropy change is thus related (as is almost everything) to the positivity of the relative entropy.

One can find the derivation of the relation (4.20) for the toy model of the generalized baker's map in section A.8 and multidimensional versions in section A.11.

**Exercise 4.5:** Random walk in an inverted potential.

Consider a particle in an inverted quadratic potential $V(x) = -\alpha x^2/2$ under the action of a random noise $\eta(t)$ with $\langle \eta(0)\eta(t) \rangle = \delta(t)$. This is described by the Langevin equation with $\alpha > 0$:

$$\dot{x} = \alpha x + \eta. \qquad (4.22)$$

Assume that the particle is at $x_0$ at $t = 0$.

(a) Find the probability distribution $\rho(x, t)$ by directly solving (4.22). Find the longtime decay of probability at a finite distance.

(b) Write the Fokker-Planck Hamiltonian $H_{FP}$. Find the spectrum of the Hamiltonian and compare it with the cases of negative and zero $\alpha$. In our case of positive $\alpha$, relate the longtime asymptotic of $\rho(x, t)$ to the lowest eigenvalue of the Fokker-Planck Hamiltonian.

# 5

# Inevitability of Irreversibility

Time is greater than space. Space is a thing.
Time, in essence, is the thought of a thing.

—JOSEPH BRODSKY

Let us understand how entropy actually grows in the physical world. A random walk increases entropy because it adds uncertainty at every step. But for a physical system, every step is prescribed by physical laws. The puzzle here is how irreversible entropy growth appears out of reversible laws of mechanics, electromagnetism, etc. If we screen the movie of any evolution backward, it will be a legitimate solution to the equations of motion. Will it have its entropy decreasing?

This question was already posed in the nineteenth century. It took the better part of the twentieth century to answer it, resolve the puzzles, and make statistical physics conceptually trivial (and technically much more powerful). The general idea is that only full knowledge can persist; any partial knowledge dissipates. Knowledge can be partial due to the inability to observe all the degrees of freedom or due to a finite precision requiring us to consider regions in phase space. The former case corresponds to Boltzmann kinetics described in section 5.2. The latter relates to the mechanism of randomization called dynamical chaos: initially small regions spread over the whole phase space under reversible Hamiltonian dynamics, very much like flows of an incompressible liquid mixing. Such spreading and mixing in phase space correspond to the approach to equilibrium, as described in section 5.3. On the contrary, to deviate a system from equilibrium, one adds external forcing and dissipation, which makes its phase flow compressible and distribution nonuniform, as described in section 5.4. In the last section, we design our own way to irreversibly forget what we consider irrelevant and learn what is relevant.

## 5.1   Evolution in the Phase Space

So far, we have said precious little about how physical systems actually evolve to arrive at equilibrium. Let us start with a broad class of energy-conserving systems that Hamiltonian dynamics can describe. Every such system is characterized by its momenta $p$ and coordinates $q$, together comprising the phase space. Any state of a system is a point in the space. Coordinates and momenta change as time progresses, and the point moves in the phase space. We should consider finite regions since we cannot measure $p, q$ exactly. We define the probability for a system to be in some $\Delta p \Delta q$ region of the phase space as the fraction $\Delta t$ of the total observation time $T$ it spends there: $w = \Delta t / T$. Assuming that the probability of finding it within the volume $dpdq$ is proportional to this volume, we introduce the statistical distribution in the phase space as a density: $dw = \rho(p, q)dpdq$. By definition, the average with the statistical distribution is equivalent to the time average:

$$\bar{f} = \int f(p, q)\rho(p, q)dpdq = \frac{1}{T}\int_0^T f(t)dt. \tag{5.1}$$

We can now consider the evolution of the density $\rho(p, q)$ on timescales larger than the $T$ used to define it.

Here we start considering flows, which are determined by the velocity **v**. Our focus is on density changes. They are brought by the flow nonuniformity, which we characterize by the velocity spatial derivatives. Consider for illustration a square with small sides, $\delta x, \delta y$, in a two-dimensional flow, $\mathbf{v}(x, y) = (v_x, v_y)$. The sides change according to $d\delta x/dt = \delta v_x = \delta x \partial v_x/\partial x$ and $d\delta y/dt = \delta v_y = \delta y \partial v_y/\partial y$. The area time derivative is as follows:

$$\frac{d}{dt}\delta x \delta y = \delta x \frac{d\delta y}{dt} + \delta y \frac{d\delta x}{dt} = \delta x \delta y \left(\frac{\partial v_x}{\partial x} + \frac{\partial v_y}{\partial y}\right) = \delta x \delta y \ div \ \mathbf{v}.$$

We see that $div \ \mathbf{v}$ gives the local rate of the volume change. Similarly, the divergence of the mass current, $\mathbf{j} = \rho \mathbf{v} = (j_x, j_y)$, determines the density change. Indeed, the differences between mass flows through the opposite $y$ sides is $\delta y \delta j_x = \delta y \delta x \partial j_x/\partial x$. Adding the difference for $x$ sides, we obtain the rate of mass change as $\delta x \delta y \ div \ \mathbf{j}$. That means that the density changes according to the continuity equation

$$\frac{\partial \rho}{\partial t} = -div\, \mathbf{j} = -div\,(\rho \mathbf{v}).$$

The phase-space flow has the velocity $\mathbf{v} = (\dot{p}, \dot{q})$. Hamiltonian dynamics of the momenta and coordinates describe the motion: $\dot{q}_i = \partial \mathcal{H}/\partial p_i$ and $\dot{p}_i = -\partial \mathcal{H}/\partial q_i$. The resulting continuity equation for the probability density is called the Liouville equation:

$$\frac{\partial \rho}{\partial t} = -div\,(\rho \mathbf{v}) = \sum_i \frac{\partial \mathcal{H}}{\partial p_i}\frac{\partial \rho}{\partial q_i} - \frac{\partial \mathcal{H}}{\partial q_i}\frac{\partial \rho}{\partial p_i} \equiv \{\rho, \mathcal{H}\}. \qquad (5.2)$$

Here the Hamiltonian generally depends on the momenta and coordinates of the given subsystem and its neighbors.

The equation $(5.2)$ describes the density evolution at a given point of the phase space since the time derivative at the left is partial, that is, taken at fixed $p_i, q_i$. Any given physical system changes its momenta and coordinates moving in the phase space. The density change for a system is then described by the full derivative taken along the flow: $d\rho/dt = \partial \rho/\partial t + (\mathbf{v}\nabla)\rho$. What is most important for us now is that any Hamiltonian flow in the phase space is incompressible: it conserves area in each plane $p_i, q_i$ and the total volume: $div\,\mathbf{v} = \partial \dot{q}_i/\partial q_i + \partial \dot{p}_i/\partial p_i = 0$. That gives the Liouville theorem: $d\rho/dt = \partial \rho/\partial t + (\mathbf{v}\nabla)\rho = -\rho\, div\,\mathbf{v} = 0$. The statistical distribution is thus conserved along the phase trajectories of any system. As a result, $\rho$ is an integral of motion.

We define statistical equilibrium as a state where $\rho$ must be expressed solely via the integrals of motion. When forces are short-range, macroscopic subsystems interact weakly and are statistically independent so that the distribution for a composite system $\rho_{12}$ is factorized: $\rho_{12} = \rho_1 \rho_2$. Since $\ln \rho_{12} = \ln \rho_1 + \ln \rho_2$ is an additive quantity, then in equilibrium it must be expressed linearly via the additive integrals of motion (which replace the enormous microscopic information). Considering a subsystem that has zero total momentum and angular momentum, the only such integral is energy $E(p, q)$, which is additive, neglecting interaction energy between subsystems. That corresponds to the familiar Gibbs canonical distribution:

$$\rho(p, q) = A \exp[-E(p, q)/T]. \qquad (5.3)$$

Note one subtlety: On the one hand, we consider weakly interacting subsystems to have their energies additive and distributions independent. On the other hand, it is precisely this weak interaction that is expected to drive a complicated evolution visiting all regions of the phase space, thus making statistical description possible. A particular case of (5.3) is a distribution constant over all the phase space (kind of microcanonical), which is evidently invariant under the Hamiltonian evolution of an isolated system due to the Liouville theorem. That distribution formally corresponds to an infinite temperature when canonical and microcanonical distributions coincide since energy differences between different regions of the phase space do not matter.

It is time for reflection. How can the Hamiltonian dynamics preserving distribution bring a system to the equilibrium distribution?

## 5.2   Kinetic Equation and *H*-Theorem

Any Hamiltonian evolution is an incompressible flow in the phase space, $\operatorname{div} \mathbf{v} = 0$, so it conserves the *total* Gibbs entropy:

$$\frac{dS}{dt} = -\int d\mathbf{x} \frac{\partial \rho}{\partial t} \ln \rho = \int d\mathbf{x} \ln \rho \operatorname{div} \rho \mathbf{v} = -\int d\mathbf{x} \, (\mathbf{v}\nabla)\rho$$

$$= \int d\mathbf{x} \, \rho \operatorname{div} \mathbf{v} = 0.$$

Which entropy then can grow? Boltzmann answered this question by deriving the equation on the one-particle momentum probability distribution. Such an equation must follow from integrating the $N$-particle Liouville equation (5.2) over all $N$ coordinates and $N-1$ momenta.

Consider the probability density $\rho(\mathbf{x}, t)$ in the phase space $\mathbf{x} = (\mathbf{P}, \mathbf{Q})$, where $\mathbf{P} = \{\mathbf{p}_1 \ldots \mathbf{p}_N\}$ and $\mathbf{Q} = \{\mathbf{q}_1 \ldots \mathbf{q}_N\}$. The Hamiltonian of particles with pair interaction is the sum of kinetic and potential energies: $\mathcal{H} = \sum_i \frac{p_i^2}{2m} + \sum_{i<j} U(\mathbf{q}_i - \mathbf{q}_j)$. The evolution of the density is described by the following Liouville equation:

$$\frac{\partial \rho(\mathbf{P}, \mathbf{Q}, t)}{\partial t} = \{\rho(\mathbf{P}, \mathbf{Q}, t), \mathcal{H}\} = \left[ -\sum_i^N \frac{\mathbf{p}_i}{m} \frac{\partial}{\partial \mathbf{q}_i} + \sum_{i<j} \theta_{ij} \right] \rho(\mathbf{P}, \mathbf{Q}, t),$$

$$(5.4)$$

where

$$\theta_{ij} = \theta(\mathbf{q}_i, \mathbf{p}_i, \mathbf{q}_j, \mathbf{p}_j) = \frac{\partial U(\mathbf{q}_i - \mathbf{q}_j)}{\partial \mathbf{q}_i} \left( \frac{\partial}{\partial \mathbf{p}_i} - \frac{\partial}{\partial \mathbf{p}_j} \right)$$

is the rate of the momentum change due to interaction. For a reduced description of the single-particle distribution over momenta, $\rho(\mathbf{p}, t) = \int \rho(\mathbf{P}, \mathbf{Q}, t)\delta(\mathbf{p}_1 - \mathbf{p})\, d\mathbf{p}_1 \ldots d\mathbf{p}_N d\mathbf{q}_1 \ldots d\mathbf{q}_N$, we integrate (5.4). The terms with $\partial/\partial\mathbf{q}_i$ do not contribute, and we get

$$\frac{\partial\rho(\mathbf{p}, t)}{\partial t} = \int \delta(\mathbf{p}_1 - \mathbf{p})\theta(\mathbf{q}_1, \mathbf{p}_1; \mathbf{q}_2, \mathbf{p}_2)\rho(\mathbf{q}_1, \mathbf{p}_1; \mathbf{q}_2, \mathbf{p}_2)\, d\mathbf{q}_1\, d\mathbf{p}_1\, d\mathbf{q}_2\, d\mathbf{p}_2.$$

(5.5)

This equation is apparently not closed since the rhs contains two-particle probability distribution. If we write the equation on that two-particle distribution integrating the Liouville equation over $N - 2$ coordinates and momenta, the interaction $\theta$ term gives three-particle distribution, etc. The consistent procedure is to consider a short-range interaction and a low density so that the mean distance between particles greatly exceeds the radius of interaction. In this case, we may assume that particles come from large distances and their momenta are not correlated for every binary collision. Statistical independence then allows us to replace the two-particle momenta distribution with the product of one-particle distributions.

It is easy to write the general form that such a closed equation must have. For a dilute gas, only two-particle collisions need to be considered in describing the evolution of the single-particle distribution over moments $\rho(\mathbf{p}, t)$. Consider the collision of two particles having momenta $\mathbf{p}, \mathbf{p}_1$:

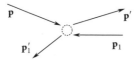

For that, they must come to the same place, yet we shall *assume* that the particle velocity is independent of the position and that the momenta of two particles are statistically independent so that the probability is the product of single-particle probabilities: $\rho(\mathbf{p}, \mathbf{p}_1) = \rho(\mathbf{p})\rho(\mathbf{p}_1)$. These strong assumptions constitute what is called the *hypothesis of molecular chaos*. Under such assumptions, the number of collisions (per unit time per unit volume) is proportional to the probabilities $\rho(\mathbf{p})\rho(\mathbf{p}_1)$ and depends on the initial momenta, $\mathbf{p}, \mathbf{p}_1$, and the final ones, $\mathbf{p}', \mathbf{p}_1'$:

$$w(\mathbf{p}, \mathbf{p}_1; \mathbf{p}', \mathbf{p}_1')\rho(\mathbf{p})\rho(\mathbf{p}_1)\, d\mathbf{p}\, d\mathbf{p}_1\, d\mathbf{p}'\, d\mathbf{p}_1'.$$

(5.6)

One may *believe* that (5.6) must work well when the one-particle distribution function evolves on a timescale much longer than that of a single collision.

We can now write the rate of the probability change as the difference between the number of particles arriving and leaving the given region of phase space around $\mathbf{p}$ by integrating over all $\mathbf{p}_1 \mathbf{p}' \mathbf{p}'_1$:

$$\frac{\partial \rho}{\partial t} = \int \left( w' \rho' \rho'_1 - w \rho \rho_1 \right) d\mathbf{p}_1 d\mathbf{p}' d\mathbf{p}'_1. \tag{5.7}$$

The scattering probabilities $w \equiv w(\mathbf{p}, \mathbf{p}_1; \mathbf{p}', \mathbf{p}'_1)$ and $w' \equiv w(\mathbf{p}', \mathbf{p}'_1; \mathbf{p}, \mathbf{p}_1)$ are nonzero only for quartets satisfying the conservation of energy and momentum. We assume that the probabilities are invariant under time reversal, which changes $\mathbf{p} \to -\mathbf{p}$ and interchanges incoming and outgoing particles:

$$w(\mathbf{p}, \mathbf{p}_1; \mathbf{p}', \mathbf{p}'_1) = w(-\mathbf{p}', -\mathbf{p}'_1; -\mathbf{p}, -\mathbf{p}_1). \tag{5.8}$$

If the medium is also invariant with respect to inversion, $\mathbf{r}, \mathbf{p} \to -\mathbf{r}, -\mathbf{p}$, then $w(\mathbf{p}, \mathbf{p}_1; \mathbf{p}', \mathbf{p}'_1) = w(-\mathbf{p}, -\mathbf{p}_1; -\mathbf{p}', -\mathbf{p}'_1)$. Translation invariance makes scattering the same at $\mathbf{r}$ and $-\mathbf{r}$. All three symmetries combined give

$$w \equiv w(\mathbf{p}, \mathbf{p}_1; \mathbf{p}', \mathbf{p}'_1) = w(\mathbf{p}', \mathbf{p}'_1; \mathbf{p}, \mathbf{p}_1) \equiv w'. \tag{5.9}$$

Using (5.9), we transform the second term in (5.7) and obtain the famous *Boltzmann kinetic equation* (1872):

$$\frac{\partial \rho}{\partial t} = \int w' \left( \rho' \rho'_1 - \rho \rho_1 \right) d\mathbf{p}_1 d\mathbf{p}' d\mathbf{p}'_1 \equiv I. \tag{5.10}$$

Actual derivation relating $w'$ in (5.10) to the interparticle potential $U$ is cumbersome; fortunately, we need only the positivity of $w'$ for what follows.

**H-*theorem*** Let us now look at the evolution of the entropy of the one-particle distribution satisfying (5.10):

$$\frac{dS}{dt} = -\int \frac{\partial \rho}{\partial t} \ln \rho \, d\mathbf{p} = -\int I \ln \rho \, d\mathbf{p}. \tag{5.11}$$

The integral (5.11) contains the integrations over all momenta so that we may exploit two interchanges, $\mathbf{p}_1 \leftrightarrow \mathbf{p}$ and $\mathbf{p}, \mathbf{p}_1 \leftrightarrow \mathbf{p}', \mathbf{p}'_1$:

$$\frac{dS}{dt} = \int w' \left( \rho \rho_1 - \rho' \rho'_1 \right) \ln \rho \, d\mathbf{p} d\mathbf{p}_1 d\mathbf{p}' d\mathbf{p}'_1$$

$$= \frac{1}{2} \int w' \left( \rho\rho_1 - \rho'\rho_1' \right) \ln(\rho\rho_1) \, d\mathbf{p} d\mathbf{p}_1 d\mathbf{p}' d\mathbf{p}_1'$$

$$= \frac{1}{2} \int w' \rho\rho_1 \ln \frac{\rho\rho_1}{\rho'\rho_1'} \, d\mathbf{p} d\mathbf{p}_1 d\mathbf{p}' d\mathbf{p}_1' \geq 0. \tag{5.12}$$

Here we subtract the integral $\int w' \left( \rho\rho_1 - \rho'\rho_1' \right) d\mathbf{p} d\mathbf{p}_1 d\mathbf{p}' d\mathbf{p}_1/2 = 0$ and use the inequality $x \ln x - x + 1 \geq 0$ with $x = \rho\rho_1/\rho'\rho_1'$. Even though the scattering probabilities are reversible in time, according to (5.8), our use of the molecular chaos hypothesis makes the kinetic equation irreversible.

Equilibrium realizes the entropy maximum, so the distribution must be a steady solution of the Boltzmann equation. Indeed, the collision integral turns into zero by virtue of $\rho_0(\mathbf{p})\rho_0(\mathbf{p}_1) = \rho_0(\mathbf{p}')\rho_0(\mathbf{p}_1')$, since $\ln \rho_0$ is the linear function of the integrals of motion, as explained in section 5.1. All this is also true for an inhomogeneous equilibrium in an external potential (see section 4.4).

One can look at the transition from (5.4) to (5.10) from a temporal viewpoint. $N$-particle distribution changes during every collision when particles exchange momenta. On the other hand, the single-particle distribution is the average over $N - 1$ particles, so changing it requires many collisions. Even though some of these collisions occur in parallel, in a dilute system with short-range interaction, the time between collisions is much longer than the collision time, so the single-particle distribution changes on a much longer scale. In other words, the transition from (5.4) to (5.10) is from a fast-changing function to a slow-changing one.

Let us summarize the present state of confusion. The full entropy of the $N$-particle distribution is conserved. The one-particle entropy grows. Is there a contradiction here? Isn't the full entropy a sum of one-particle entropies? The answer (no to both questions) follows from our consideration of mutual information in section 2.8. What was defined as the entropy of the gas in thermodynamics is indeed the sum of entropies of different particles $\sum S(p_i, q_i)$. In the thermodynamic limit, we neglect interparticle correlations. However, the total entropy of the gas includes correlations, which are measured by generalized (multiparticle) mutual information:

$$S(p_1 \ldots p_n, q_1, \ldots q_n) = \sum_i S(p_i, q_i) - I(p_1, q_1; \ldots; p_n, q_n).$$

We broke time reversibility and set the arrow of time when we assumed particles were uncorrelated before the collision and not after. If one starts from a set of uncorrelated particles and lets them interact, then the interaction will build correlations, and the total distribution will change, but the total entropy will not. This is because lowering entropy by correlations compensates the growth of the single-particle entropies. That growth is described by the Boltzmann equation, which is valid for an uncorrelated initial state (and for some time after). The motivation for choosing such an initial state for computing one-particle evolution is that it is most likely in any generic ensemble. Yet it would make no sense to run the Boltzmann equation backward from a correlated state, which is statistically a very unlikely initial state, since it requires momenta to be correlated in such a way that a definite state is produced after time $t$. In other words, the Boltzmann equation describes at a macroscopic level (of one-particle distribution) not all but typical microscopic ($N$-particle) evolutions.

We can replace the usual second law of thermodynamics by the law of conservation of the total entropy (or information): the increase in the thermodynamic (uncorrelated) entropy is exactly compensated by the increase in correlations between particles expressed by the mutual information. The usual second law results from our renunciation of all correlation knowledge and not from any intrinsic behavior of dynamical systems. One way to disregard correlations is to consider only one-particle distribution as we do here. Another version of such renunciation is presented in section 5.3: the full $N$-particle entropy grows because of phase-space mixing and continuous coarse-graining.

But mutual information can work in the opposite direction, too. Imagine that two systems are at respectively $T_1$ and $T_2$, heat $dE_1$ passes from 2 to 1, and the degree of their correlation changes by $\Delta I$. The second law then generalizes (1.6) to

$$\left( \frac{1}{T_1} - \frac{1}{T_2} \right) dE_1 - \Delta I \geq 0. \tag{5.13}$$

If correlations were absent before and appeared when the systems were brought into contact, then $\Delta I > 0$ and we still have heat flowing from hot to cold, its amount bounded from below: $dE_1(T_2 - T_1) \geq T_1 T_2 \Delta I > 0$. However, one can create a situation where there is an initial correlation between the systems that is destroyed during the heat exchange, i.e., $\Delta I < 0$. In this case, the heat could flow from the cold to the hot system. An information-theoretic resource can be used to perform refrigeration using, for instance, Maxwell's demon, who opens a window for fast particles from the right and slow particles from the left.

Neglecting interparticle correlations by factorizing the two-particle distribution $\rho_{12} = \rho(\mathbf{q_1}, \mathbf{p_1}; \mathbf{q_2}, \mathbf{p_2}) = \rho_1 \rho_2$ means using incomplete information. This naturally leads to a further increase of uncertainty, that is, of entropy. For dilute gases, such a factorization is just the first term of an expansion over the powers of density:

$$\rho_{12} = \rho_1 \rho_2 + \int d\mathbf{q_3} d\mathbf{p_3} J_{123} \rho_1 \rho_2 \rho_3 + \dots .$$

In section A.7, we explain that this (so-called cluster) expansion is regular only for equilibrium distributions. For nonequilibrium distributions, the expansion generally contains nonanalytic terms with $\log \rho$. The Boltzmann equation is nice, but corrections to it are ugly when one deviates from equilibrium. The corrections also violate the $H$-theorem—indeed, dropping all the terms is part of passing from the Liouville equation to the Boltzmann equation, which leads to the loss of information and entropy growth.

## 5.3   Phase-Space Mixing and Entropy Growth

We have seen that one-particle entropy can grow even when the full $N$-particle entropy is conserved. To show how the full entropy can grow, let us return to the full $N$-particle distribution and recall that we always measure coordinates and momenta within some intervals; i.e., we characterize the system not by a point but by a finite region in phase space. We now show that quite general dynamics stretches this finite domain into a thin, convoluted strip whose parts can be found everywhere in the available phase space, say, on a fixed-energy surface. The dynamics thus provide a stochastic-like mixing in phase space responsible for the approach to equilibrium (uniform microcanonical distribution). By itself, this stretching and mixing does not change the phase volume and entropy. Another necessary ingredient is to continually treat our system with finite precision. Such a consideration is called *coarse-graining*, and together with mixing, it is responsible for the irreversibility of statistical laws and for entropy growth.

The dynamical mechanism of entropy growth is the separation of trajectories in phase space: trajectories started from a small neighborhood are found farther and farther away from each other as time proceeds. Denote again by $\mathbf{x} = (\mathbf{P}, \mathbf{Q})$ the $6N$-dimensional vector of the position and by $\mathbf{v} = (\dot{\mathbf{P}}, \dot{\mathbf{Q}})$ the velocity in the phase space. The relative motion of two close points, separated by $\mathbf{r}$, is determined by their velocity difference: $\delta v_i \approx r_j \partial v_i / \partial x_j = r_j \sigma_{ij}$. We have seen in section 5.1 that the trace of the tensor of velocity derivatives, $div\ \mathbf{v} = \sum_i \sigma_{ii}$, determines the volume change rate. We now consider the whole tensor and decompose it into an antisymmetric part (describing rotation) and a symmetric part, $S_{ij} = (\partial v_i/\partial x_j + \partial v_j/\partial x_i)/2$ (describing deformation). Separation of trajectories is due to deformation, so we focus on $S_{ij}$. The vector initially parallel to the axis $j$ turns toward the axis $i$ with the angular speed $\partial v_i/\partial x_j$, so that $2S_{ij}$ is the rate of change of the angle between two mutually perpendicular vectors along $i$ and $j$ axes. To put it simply, $2S_{ij}$ is the rate at which a rectangle deforms into a parallelogram. Arrows in the figure show the velocities of the endpoints:

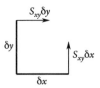

The symmetric tensor $S_{ij}$ can be always transformed into a diagonal form by an orthogonal transformation (i.e., by the rotation of the axes) so that $S_{ij} = S_i\delta_{ij}$. According to the Liouville theorem, Hamiltonian dynamics is an incompressible flow in the phase space, so the trace must be zero: $\mathrm{Tr}\,\sigma_{ij} = \sum_i S_i = \mathrm{div}\,\mathbf{v} = 0$. That means that some components are positive and some are negative. Positive diagonal components are the stretching rates, and negative components are the contraction rates in respective directions. The equation for the distance between two points along a principal direction has the form $\dot{r}_i = \delta v_i = r_i S_i$. The solution is as follows:

$$r_i(t) = r_i(0) \exp\left[\int_0^t S_i(t')\,dt'\right]. \tag{5.14}$$

For a time-independent strain, the growth/decay is exponential over time.

A purely straining motion converts a spherical element into an ellipsoid with the principal diameters that grow or decay. As an example, consider a two-dimensional projection of the initial spherical element, i.e., a circle of the radius $R$ at $t = 0$, as shown in figure 5.1. The point that starts at $x_0, y_0 = \sqrt{R^2 - x_0^2}$ becomes

$$x(t) = e^{S_{11}t}x_0,$$

$$y(t) = e^{S_{22}t}y_0 = e^{S_{22}t}\sqrt{R^2 - x_0^2} = e^{S_{22}t}\sqrt{R^2 - x^2(t)e^{-2S_{11}t}},$$

$$x^2(t)e^{-2S_{11}t} + y^2(t)e^{-2S_{22}t} = R^2. \tag{5.15}$$

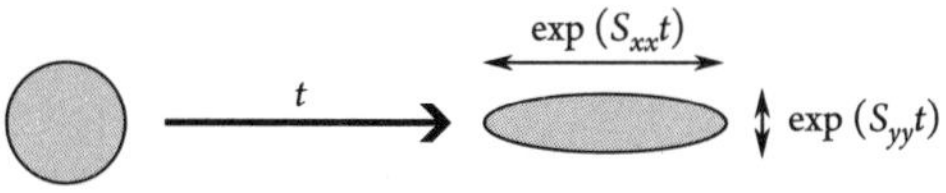

FIGURE 5.1. Deformation of a phase-space element by a permanent strain.

The equation (5.15) describes how the initial circle turns into the ellipse whose eccentricity increases exponentially with the rate $|S_{11} - S_{22}|$. In a multidimensional space, any sphere of initial conditions turns into the ellipsoid defined by

$$\sum_{i=1}^{6N} x_i^2(0) = \sum_{i=1}^{6N} x_i^2(t) e^{-2S_i t} = \text{const.}$$

If our uncertainty about the initial state is confined within a sphere, then the uncertainty about the evolved state is within the ellipsoid. As the system moves in the phase space, both the strain values and the orientation of the principal directions change so that an expanding direction may turn into a contracting one and vice versa. Since we do not want to go into the details of dynamics, we consider such evolution as a kind of random process. The question is whether averaging over all values and orientations gives a zero net separation of trajectories. It may seem counterintuitive, but exponential stretching generally persists on average, and the majority of trajectories separate. There are two ways to understand that: one in space and another in time.[1]

Let us first go with the flow and see the separation of trajectories with time. Denote the rate of separation along a given direction as $\Lambda_i(t) = \int_0^t S_i(t')dt'$. Even when the time average is zero, $\lim_{t\to\infty} \int_0^t S_i(t')dt' = 0$, its average exponent is larger than unity (and generally grows with time):

$$\left\langle \left| \frac{r_i(t)}{r_i(0)} \right| \right\rangle = \lim_{T\to\infty} \frac{1}{T} \int_0^T dt\, e^{\Lambda_i(t)} \geq 1. \tag{5.16}$$

This is because the time intervals with positive $\Lambda(t)$ contribute more to the exponent than the intervals with negative $\Lambda(t)$. That follows from the *convexity* of the exponential function. In the simplest case, when $\Lambda$ is uniformly distributed over the interval $-a < \Lambda < a$, the average $\Lambda$ is zero, while the average exponent exceeds unity: $(1/2a) \int_a^{-a} e^\Lambda d\Lambda = (e^a - e^{-a})/2a > 1$.

---

1. "Time and space are modes by which we think and not conditions in which we live" (A. Einstein).

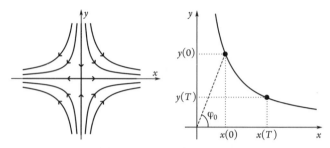

FIGURE 5.2. Left: Streamlines of a saddle-point flow. Right: Motion down a streamline. For $\varphi = \varphi_0 = \arccos[1 + \exp(2\lambda T)]^{-1/2} > \pi/4$, the initial and final points are symmetric relative to the diagonal: $x(0) = y(T)$ and $y(0) = x(T)$. If $\varphi < \varphi_0$ (majority of starting points), the distance from the origin increases.

From a spatial perspective, let us look at figure 5.2, presenting the simplest case of a pure strain, which corresponds to an incompressible saddle-point flow in a plane: $v_x = \lambda x$, $v_y = -\lambda y$. We are in the reference frame of the trajectory corresponding to the center, so that $\mathbf{r} = (x, y)$ represents the separation between that trajectory and one nearby. Two-dimensional phase-space flow is of great illustrative value, all the more because the Liouville theorem is true in every $p_i - q_i$ plane projection. Here we have one expanding direction and one contracting direction, their rates being equal. The evolution of the vector components satisfies the equations $\dot{x} = v_x$ and $\dot{y} = v_y$. The solutions, $x(t) = x_0 \exp(\lambda t)$ and $y(t) = y_0 \exp(-\lambda t) = x_0 y_0/x(t)$, show that every trajectory is a hyperbole. Whether the separation vector is stretched or contracted after some time $T$ depends on its orientation and on $T$. The vector $\mathbf{r} = (x, y)$ can initially have any angle $\varphi$ with the $x$ axis. After time $T$, the length is multiplied by $[\cos^2 \varphi \exp(2\lambda T) + \sin^2 \varphi \exp(-2\lambda T)]^{1/2}$. The vector is stretched if $\cos \varphi \geq [1 + \exp(2\lambda T)]^{-1/2} < 1/\sqrt{2}$, i.e., the fraction of stretched directions is larger than half. When all orientations are equally probable along the motion, the net effect is stretching, increasing with the persistence time $T$.

The net stretching and separation of trajectories is formally proved in mathematics by considering a random strain matrix $\hat{\sigma}(t)$ and the transfer matrix $\hat{W}$ defined by $\mathbf{r}(t) = \hat{W}(t, t_1)\mathbf{r}(t_1)$. It satisfies the equation $d\hat{W}/dt = \hat{\sigma}\hat{W}$. The Liouville theorem $\mathrm{tr}\,\hat{\sigma} = 0$ means that $\det \hat{W} = 1$. The modulus $r(t)$ of the separation vector may be expressed via the positive symmetric matrix $\hat{W}^T \hat{W}$. The main result (Furstenberg and Kesten 1960; Oseledec 1968) states that in almost every realization $\hat{\sigma}(t)$, the matrix $\frac{1}{t} \ln \hat{W}^T(t, 0)\hat{W}(t, 0)$ tends to a finite limit as $t \to \infty$. In particular, its eigenvectors tend to $d$ fixed orthonormal eigenvectors $\mathbf{f}_i$. Geometrically, that means that an initial sphere evolves into an elongated ellipsoid over time. The limiting eigenvalues

$$\lambda_i = \lim_{t \to \infty} t^{-1} \ln |\hat{W} \mathbf{f}_i| \qquad (5.17)$$

define the so-called Lyapunov exponents, which can be thought of as the mean stretching rates. The sum of the exponents is the mean volume growth rate, which is zero due to the Liouville theorem. As long as there is no special degeneracy, which makes all the exponents identically zero, at least one positive exponent gives stretching. Therefore, as time increases, the ellipsoid becomes more and more elongated, and it is less and less likely that the hierarchy of the ellipsoid axes will change.

The mathematical lesson to learn is that by multiplying $N$ random matrices with a unit determinant (recall that the determinant is the product of eigenvalues), one generally gets some eigenvalues growing and some decreasing exponentially with $N$. It is also worth remembering that there is always a probability for two trajectories to come closer in a random flow. That probability decreases with time, but it is finite for any finite time. In other words, the majority of trajectories separate, but some approach. The separating ones provide for the exponential growth of positive moments of the distance: $E(a) = \lim_{t \to \infty} t^{-1} \ln \left[ \langle r^a(t)/r^a(0) \rangle \right] > 0$ for $a > 0$. However, approaching trajectories have $r(t)$ decreasing, guaranteeing that the moments with sufficiently negative $a$ also grow. We mention without proof that $E(a)$ is a concave function that passes through zero, $E(0) = 0$. It must then have another zero, which can be shown to be $a = -d$ for isotropic random flow in $d$-dimensional space.

The probability of finding a ball turning into an exponentially stretching ellipse thus goes to unity as time increases. The physical reason for this is that substantial deformation appears sooner or later. To reverse it, one needs to contract the long axis of the ellipse. The direction of contraction then must be inside the narrow angle defined by the ellipse eccentricity, which is less likely than being outside the angle:

To transform ellipse to circle, contracting direction must be within the angle.

This is similar to the argument about the irreversibility of the Boltzmann equation in the previous section. Randomly oriented deformations continue to increase the eccentricity on average.

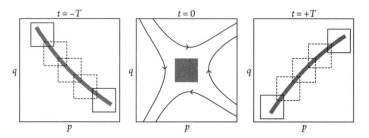

FIGURE 5.3. Increase of the phase volume upon stretching-contraction and coarse-graining. The central panel shows the initial state and the velocity field.

Armed with the understanding of exponential stretching, we now return to the dynamical foundation of the second law of thermodynamics. Our finite resolution does not allow us to distinguish between the states within some square in the phase space. That square is our "grain" in coarse-graining. In figure 5.3, one can see how such a black square of initial conditions (at the central panel) is stretched in one (unstable) direction and contracted in another (stable) direction so that it turns into a long, narrow strip (left and right panels). Later in time, our resolution is still restricted—the rectangles in the right panel show finite resolution (this is coarse-graining). Viewed with such resolution, our set of points occupies a larger phase volume at $t = \pm T$ than at $t = 0$. A larger phase volume corresponds to larger entropy.

The time reversibility of any trajectory does not contradict the time-irreversible filling of the space by the set of trajectories considered with a finite resolution. By reversing time, we exchange stable and unstable directions (i.e., those of contraction and expansion), but the fact of space filling persists. We see from figure 5.3 that the volume and entropy increase both forward and backward in time. And yet our consideration does provide for a time arrow: if we observe an evolution that produces a narrow strip, then its time reversal is the contraction into a ball; but if we consider a narrow strip as an initial condition, it is unlikely to observe a contraction because of the narrow angle mentioned above. Therefore, being shown two (sufficiently long) movies, one with stretching into a strip and another with contraction into a ball, we conclude that with probability close (but not exactly equal!) to unity, the first movie shows the true sequence of events from the past to the future.

When the possible occupied region expands, the entropy grows as the logarithms of the volume. If initially our system is within the phase-space volume $v_0$, then its density is $\rho_0 = 1/v_0$ inside and zero outside. After stretching to some larger volume $e^{\lambda t} v_0$, the entropy $S = -\int \rho \ln \rho d\mathbf{x}$ has increased by $\lambda t$. If there are $k$ stretching and $d - k$ contracting directions in a $d$-dimensional space, then contractions eventually stabilize at the resolution scale while expansions continue. Therefore, the long-term entropy growth rate is determined by the sum of the positive Lyapunov exponents, $\lambda = \sum_{i=1}^{k} \lambda_i$.

Let us briefly discuss our flow from the information perspective. Consider an ensemble of systems having close initial positions within our finite resolution. In a flow with positive Lyapunov exponents, we lose our ability to predict where it goes with time. This loss of information is determined by the growth of the available phase volume, that is, of the entropy. But we can look backward in time and ask where the points come from. If we consider two points along a stretching direction, we can with confidence predict that they were closer before. During some time in the past, they were hidden inside the resolution circle, but they separate with time beyond the resolution and can now be distinguished:

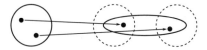

As time proceeds, we learn more and more about the initial locations of the points. The acquisition rate of such information about the past is again the sum of the positive Lyapunov exponents and is called the Kolmogorov-Sinai entropy. As time lag from the present moment increases, we can say less and less where we shall be and more and more where we came from. It illustrates Kierkegaard's remark that the irony of life is that it is lived forward but understood backward.

After the strip length reaches the scale of the velocity change (when one already cannot approximate the phase-space flow by a linear profile $\hat{\sigma} r$), the strip starts to fold because rotation (which we can neglect for a ball but not for a long strip) is different at different parts of the strip. Still, however long and folded, the strip continues the exponential stretching locally. Eventually, one can find the points from the initial ball everywhere, which means that the flow is mixing, or ergodic. A formal definition is that the flow is called *ergodic* in the domain if the trajectory of almost every point (except possibly a set of zero volume) passes arbitrarily close to every other point. An equivalent definition is that there are no finite-volume subsets of the domain invariant with respect to the flow except the domain itself. Ergodic flow on an energy surface in the phase space provides for a microcanonical distribution (i.e., constant) since time averages are equivalent to the average over the surface. While we can prove ergodicity only for relatively simple systems, like the gas of hard spheres, we believe that it holds for most systems of a sufficiently general nature. (That vague notion can be made more precise by saying that the qualitative behavior is insensitive to small variations of the system's microscopic parameters.)

One can think of Hamiltonian dynamics as a phase space mapped onto itself. Section A.8 describes a toy model of such a map, which is of great illustrative value for the applications of chaos theory to statistical mechanics.

Two concluding remarks are in order. First, the notion of an exponential separation of trajectories put an end to the old dream of Laplace to be able to predict the future if only all coordinates and momenta are given. Even if we were able to measure all relevant phase-space initial data, we could do it only with a finite precision $\epsilon$. However small the indeterminacy in the data, it is amplified exponentially with time so that eventually $\epsilon \exp(\lambda T)$ is large, and we cannot predict the outcome. Mathematically speaking, limits $\epsilon \to 0$ and $T \to \infty$ do not commute. Second, the above arguments do not use the usual mantra of the thermodynamic limit, which means that even the systems with a small number of degrees of freedom need statistics for their description at long times if their dynamics have a positive Lyapunov exponent (which is generic). This is sometimes called *dynamical chaos*.

A common lesson from the last two sections is that full knowledge persists while partial knowledge dissipates. If you know everything, this knowledge stays with you (the Liouville theorem). But if your knowledge is incomplete— either because you study only part of your degrees of freedom (Boltzmann) or because of finite precision (coarse-graining)—then your degree of uncertainty generally increases with time.

## 5.4   Entropy Decrease and
## Nonequilibrium Fractal Measures

As we have seen in the previous section, if we have indeterminacy in the data or consider an ensemble of systems, then an incompressible flow of Hamiltonian dynamics effectively mixes and makes the distribution uniform in the phase space. Evolution is Hamiltonian for isolated systems, which conserve their integrals of motion so that the distribution is uniform over the respective surfaces of constant integrals. In particular, dynamical chaos justifies microcanonical distribution, uniform over the energy surface.

To diminish entropy, one needs to act. Let us now consider systems that are not isolated; the dynamics are non-Hamiltonian, and the Liouville theorem is invalid. The flow in the phase space is then generally compressible. The simplest nonconservative effect is a dissipation of kinetic energy, which shrinks all momenta and thus decreases the phase volume. We are interested, however, in a nonequilibrium steady state where we keep the total energy nondecreasing. For example, to compensate for the loss of momentum of particles with dissipation rates $\gamma_i$, we act on them by external forces $f_i$, so that the equations of motion take the form (4.7)

$$\dot{p}_i = f_i - \gamma_i p_i - \frac{\partial H}{\partial q_i}, \quad \dot{q}_i = \frac{\partial H}{\partial p_i} \Rightarrow div\,\mathbf{v} = \sum_i \left( \frac{\partial f_i}{\partial p_i} - \gamma_i \right).$$

When the system is in a thermostat, the forces $f_i$ are due to random kicks, which are short-correlated compared to times of order $\gamma_i^{-1}$. Such forces are in detailed balance with the dissipation: after averaging over the short correlation time, $\langle \partial f_i / \partial p_i \rangle = \gamma_i$ for every $i$, so that $div\,\mathbf{v} \equiv 0$. For an example, see the consideration of a Brownian particle in section A.10, particularly (A.27).

Let us consider now a generic environment, where there is no detailed balance and the forces are correlated so that $div\,\mathbf{v} \neq 0$ during finite intervals. The whole phase volume does not change, that is, the volume integral of the local expansion rate is zero at every moment: $\int div\,\mathbf{v}\,d\mathbf{r} = 0$. Such phase-space flows create quite different distributions since the probability density changes along a flow: $d\rho/dt = -\rho\,div\,\mathbf{v}$. For a nonuniform density, the entropy is not the (Boltzmann) log of the phase volume but the (Gibbs) *mean* log of the inverse density, $S(t) = -\langle \ln \rho \rangle = -\int \rho(\mathbf{r}, t) \ln \rho(\mathbf{r}, t)\,d\mathbf{r}$. The entropy production rate equals the mean local expansion rate:

$$\frac{dS}{dt} = \int \rho(\mathbf{r}, t) div\,\mathbf{v}(\mathbf{r}, t)\,d\mathbf{r} = \langle div\,\mathbf{v} \rangle. \qquad (5.18)$$

Even though $\int div\,\mathbf{v}\,d\mathbf{r} = 0$, (5.18) is nonzero because of correlations between $\rho$ and $div\,\mathbf{v}$. Indeed, $\rho$ is on average smaller in the expanding regions where $div\,\mathbf{v} > 0$. That means that (5.18) is nonpositive and the entropy decreases. Maximal entropy corresponds to a uniform distribution. The decrease in entropy, keeping normalization $\int \rho(\mathbf{r}, t)\,d\mathbf{r} = 1$, means (by convexity) that the distribution is becoming more and more nonuniform in the phase space.

Let us now switch focus from space to time and consider the density of an arbitrary fluid element with the coordinate $\mathbf{r}(t)$, which satisfies $d\mathbf{r}/dt = \mathbf{v}$ and

$\mathbf{r}(0) = \mathbf{r}_0$. The density then evolves along the flow as follows:

$$\frac{\rho(\mathbf{r}(t), t)}{\rho(\mathbf{r}_0, 0)} = \exp\left[-\int_0^t div\, \mathbf{v}(\mathbf{r}(t'), t')\, dt'\right] = e^{C(t)}. \tag{5.19}$$

If the expansion rate in the flow reference frame, $s(t) = div\, \mathbf{v}(\mathbf{r}(t), t)$, is a random function with a finite correlation time $\tau$, then its integral $C = \int_0^t s(t')\, dt'$ at $t/\tau = N \gg 1$ can be broken into a sum of many uncorrelated random numbers with a zero mean and some variance $\Delta$. According to the central limit theorem (see section A.2), the statistics of such a sum are Gaussian with a zero mean and the variance linearly growing with time: $\mathcal{P}(C) \propto e^{-C^2/2\Delta N}$. We then obtain for the average over all possible trajectories $\overline{\exp(C)} = \int \mathcal{P}(C) e^C dC \propto e^{N\Delta/2}$. Therefore, for a generic random flow, the density of most fluid elements must grow nonstop as they move. The reason is again the concavity of the exponential function, as in (5.16): if the mean is zero, the mean exponent generally exceeds unity.

Since the total measure is conserved, the growth of density in some places must be compensated by a decrease in other places so that the distribution becomes more and more nonuniform, which decreases the entropy. Looking at the phase space, one sees it more and more emptied, with the density concentrated asymptotically in time on a small subset. That is the opposite of mixing by Hamiltonian incompressible flow. Note how arguing for the entropy decrease we used the convexity of the logarithm in the spatial consideration and the concavity of the exponent in the temporal argument.

If the density of any fluid element grows on average, its volume decreases. The longtime Lagrangian average (along the flow) of the volume change rate,

$$\overline{\frac{dS}{dt}} = \overline{div\, \mathbf{v}} = \lim_{t \to \infty} \frac{1}{t} \int_0^t div\, \mathbf{v}(t')\, dt' = \sum_i \lambda_i,$$

is a sum of the Lyapunov exponents, which is then nonpositive, in contrast to an instantaneous average over space, which is zero at any time: $\int div\, \mathbf{v}\, d\mathbf{r} = 0$.

It is important that we allowed for compressibility of a phase-space flow, $\mathbf{v}(\mathbf{r}, t)$, but did not require its irreversibility. Even if the system is invariant with respect to $t \to -t$, $\mathbf{v} \to -\mathbf{v}$, the entropy production rate is nonnegative and the sum of the Lyapunov exponents is nonpositive for the same simple reason that contracting regions have more measure and give higher contributions. Backward in time, the measure also concentrates, only on a different set.

FIGURE 5.4. The number of covering squares depends on their size: $N(1/2) = 4, N(1/4) = 12, N(1/8) = 30$.

Let us show that in a spatially smooth random compressible flow the density could concentrate on a very nonsmooth set having dimensionality less than that of the full phase space. The dimensionality could even be noninteger, which corresponds to a fractal set. One defines the (box-counting) dimension of a set as follows:

$$d_f = \lim_{\epsilon \to 0} \frac{\ln N(\epsilon)}{\ln(L/\epsilon)}. \tag{5.20}$$

Here $N(\epsilon)$ is the number of boxes of side $\epsilon$ needed to cover the set of the size $L$ (see figure 5.4).

Consider a two-dimensional phase-space flow with one positive and one negative Lyapunov exponent, $\lambda_+$ and $\lambda_-$. After time $t$, a square having initial side $\delta \ll L$ will be stretched into a long, thin strip of length $\delta \exp(t\lambda_+)$ and width $\delta \exp(t\lambda_-)$. To cover the contracting direction, we choose $\epsilon = \delta \exp(t\lambda_-)$, then $N(\epsilon) = \exp[t(\lambda_+ - \lambda_-)]$, so that the dimension is

$$d_f = 1 + \frac{\lambda_+}{|\lambda_-|}. \tag{5.21}$$

Since $|\lambda_-| \geq \lambda_+$, the dimension is between 1 and 2. The set is smooth in the expanding direction and fractal in the contracting direction, giving two terms in (5.21). How density concentrates on a fractal set in a random compressible flow is illustrated by a toy model presented in section A.8.

The general (Kaplan-Yorke) conjecture is that $d_f = j + \sum_{i=1}^{j} \lambda_i/\lambda_{j+1}$, where $j$ is the largest number for which $\sum_{i=1}^{j} \lambda_i \geq 0$ and $\sum_{i=1}^{j+1} \lambda_i < 0$. For incompressible flows, $j = d$.

Fractalization of the measure proceeds until the coarse-graining stops it. In contrast to the incompressible flow, coarse-graining at a small scale $\epsilon$ does not make the distribution uniform, but it makes the entropy finite: $S = \ln N(\epsilon) = d_f \ln(L/\epsilon)$. An equilibrium uniform (microcanonical) distribution in $d$-dimensional phase space has the entropy $S_0 = d \ln(L/\epsilon)$; a nonequilibrium steady state generally has a lower dimensionality, $d_f < d$, with a lower entropy.

We thus see that, for smooth dynamical systems, both temporal and spatial properties of the entropy are determined by the Lyapunov exponents. Entropy dependence on time (both forward and backward) is governed by the Kolmogorov-Sinai entropy, which is the sum of the positive Lyapunov exponents. The dimensionality determines entropy dependence on spatial resolution.

Let us appreciate the dramatic difference between the entropy growth described in section 5.3 and the entropy decay described in the present section (see also section A.8 for examples of both). In the former, phase-space flows were area-preserving and the volume growth of an element was due to a finite resolution, which stabilized the size in the contracting direction so that the mean volume growth rate was solely due to stretching directions and thus equal to the sum of the positive Lyapunov exponents, as described in section 5.3. On the contrary, the present section deals with compressible flows. The relation between compressibility and nonequilibrium is natural: to make a system non-Hamiltonian, one needs to pump energy into some degrees of freedom and absorb it from other degrees of freedom to keep a steady state, which corresponds to expansion and contraction of the momentum part of the phase space. That decreases entropy by creating more inhomogeneous distributions. The mean rate of the entropy decay is the sum of all the Lyapunov exponents, which is nonpositive since contracting regions contain more trajectories and contribute more than expanding regions. Long time net contraction of a fluid element and respective entropy decay is the analog of the second law of thermodynamics: to deviate a system from equilibrium, one needs to lower its entropy until the resolution limit is reached.

This is a good time to reflect on the complementarity of determinism and randomness expressed in terms "statistical mechanics" (nineteenth century) and "dynamical chaos" (twentieth century). What shall we have in the twenty-first century: predictable uncertainty, multiversion reality?

## 5.5   Renormalization Group and the Art of Forgetting

Erase the features Chance installed, and you will see the world's great beauty.

—ALEXANDER BLOK

Erase the features Chance installed. Watch by chance do not rub a hole.

—VSEVOLOD NEKRASOV[2]

Economics, biology, and physics all deal with what are essentially large-scale, low-resolution effective theories. Even what were once considered elementary particles are now described as large-scale excitations of fields whose microscopic behavior is generally unknown (say, at the Planck scale, introduced in section 6.6). The most fundamental question is which information about the microscopic properties determines the observable macroscopic behavior, and which is irrelevant and can be forgotten. We have seen in this chapter how dynamics naturally lose information. Rather than leave it to Nature, we ourselves can design the process of forgetting. One such step-by-step process is called the renormalization group (RG). It eliminates degrees of freedom, renormalizes remaining ones, and looks for universal features of the statistical distributions that are invariant with respect to such a procedure. There is a paradigm shift brought by the renormalization group approach. Instead of being interested in this or that probability distribution, we are interested in different RG flows in the space of distributions. Under RG transformation, whole families (universality classes) of systems described by different distributions flow to the same fixed point (i.e., have the same asymptotic distribution).

As with almost everything in this book, the simplest realization of RG sums independent random numbers, a procedure described in detail in Section A.2. Let us do it step-by-step, summing pairs at every step. Consider a set of random independent, identically distributed (iid) variables $\{x_1 \ldots x_N\}$, each having the probability distribution $\rho(x)$ with zero mean and unit variance. The two-step RG reduces the number of random variables by replacing any two of them by their sum and rescales the sum to keep the variance: $z_i = (x_{2i-1} + x_{2i})/\sqrt{2}$. Since summing doubles the variance, we divide by $\sqrt{2}$.

---

2. Translated from Russian by A. Shafarenko.

Each new random variable has the following distribution:

$$\rho'(z) = \int dx dy \rho(x)\rho(y)\delta\left(z - \frac{x+y}{\sqrt{2}}\right). \tag{5.22}$$

The distribution, which does not change under the procedure, is called a fixed point (in the space of functions) and satisfies the equation

$$\rho(x) = \sqrt{2}\int dy \rho(y)\rho(\sqrt{2}x - y).$$

Since this is a convolution equation, we solve it by the Fourier transform, $\rho(k) = \int \rho(x)e^{ikx}dx$. Multiplying by $e^{ikx\sqrt{2}}$ and integrating, we get

$$\rho(k\sqrt{2}) = \rho^2(k). \tag{5.23}$$

The solution of (5.23) is $\rho_0(k) \sim e^{-k^2}$ and $\rho_0(x) = (2\pi)^{-1/2}e^{-x^2/2}$. We thus have shown that the Gaussian distribution is a fixed point of repetitive summation and rescaling of random variables, keeping variance fixed. This is not surprising, since it has a maximal entropy among the distributions with the same variance.

To turn that into the central limit theorem, we need to show that this distribution is stable, that is, RG flows toward it. For the flow near the fixed point, we denote $\rho = \rho_0(1+h)$ and linearize the transform in $h$. The transformed distribution is then $h'(k) = 2h(k/\sqrt{2})$. The eigenfunctions of the linearized transform are $h_m(k) = k^m$ with eigenvalues $h'_m(k)/h_m(k) = 2^{1-m/2}$. We see that the modes with $m = 0, 1$ grow, while the mode with $m = 2$ does not decay. Fortunately, these three modes are forbidden by the three conservation laws of the transformation (5.22): the moments $\int x^n \rho(x)\,dx$ must be preserved for $n = 0$ (normalization), $n = 1$ (zero mean), and $n = 2$ (unit variance). The moments of $\rho(x)$ are the derivatives of the generating function $\rho(k)$ at $k = 0$: $\int x^n \rho(x)\,dx = d^n \rho(k)/d(ik)^n_{k=0}$. Therefore, the three conservation laws mean that $h(0) = dh(0)/dk = d^2h(0)/dk^2 = 0$, so only the perturbation modes with $m > 2$ are admissible. All the admissible perturbations decay upon RG flow, that is, deviations from the fixed point decrease, which means that the point is an attractor.

To conclude, the RG flow eventually brings us to the distribution with the maximal entropy, forgetting all the information except the invariants— normalization, the mean, and the variance.

When we look for limiting distributions in the real world, we often need to deal not with independent but with strongly correlated random variables. Let

Rescaling by predictive mind.

us consider the Ising model of interacting spins and describe RG as the procedure of block spin transformation. The model is mentioned in section 3.1: random variables are spins, $\sigma_i = \pm 1$. To eliminate small-scale degrees of freedom, we divide all the spins into groups (blocks). It is natural to group into blocks the most strongly correlated spins. In neuron systems (3.3), correlation is not necessarily related to spatial proximity. Here we consider physical systems where the strongest correlations are with the nearest neighbors. In this case, there are $m^d$ spins in every block with the side $m$ ($d$ is space dimensionality). We then assign to any block a new variable $\sigma'$, which is $\pm 1$ when the spins in the block are predominantly up or down. We *assume* that the system can be described equally well in terms of block spins, with the distribution of the same form as the original but with renormalized parameters.

Consider first a one-dimensional chain, where the Gibbs distribution is

$$\rho\{\sigma_i\} = Z^{-1} \exp\left(-K \sum_i \sigma_i \sigma_{i+1}\right). \tag{5.24}$$

It has a single parameter, $K = 1/T$, which will be renormalized. The partition function $Z$ is easy to compute by summing not over $N$ spins but over the $N-1$

bonds between them. A bond brings either factor $e^K$ when two spins have the same sign or $e^{-K}$ when the signs are different. For a chain with open ends, we also have two possible values at the ends, which gives

$$Z(K) = \sum_{\{\sigma = \pm 1\}} \exp\left[K \sum_i \sigma_i \sigma_{i+1}\right] = 2(2 \cosh K)^{N-1}.$$

Let us transform $\rho\{\sigma_i\}$ and $Z(K)$ by the procedure (called decimation[3]) of eliminating degrees of freedom by ascribing (undemocratically) to every block of $m = 3$ spins the value of the central spin. Consider two neighboring blocks, $\sigma_1, \sigma_2, \sigma_3$ and $\sigma_4, \sigma_5, \sigma_6$, and sum over all values of $\sigma_3 = \pm 1$, $\sigma_4 = \pm 1$, keeping $\sigma_1' = \sigma_2$ and $\sigma_2' = \sigma_5$ fixed. The respective factors in the partition function can be written as follows: $\exp[K\sigma_3\sigma_4] = \cosh K + \sigma_3\sigma_4 \sinh K$, which is true for $\sigma_3\sigma_4 = \pm 1$. Denote $x = \tanh K$. Then only the terms with even powers of $\sigma_3$ and $\sigma_4$ contribute the factors in the partition function that involve these degrees of freedom:

$$\sum_{\sigma_3, \sigma_4} e^{K(\sigma_1'\sigma_3 + \sigma_3\sigma_4 + \sigma_4\sigma_2')} = \cosh^3 K \sum_{\sigma_3, \sigma_4} (1 + x\sigma_1'\sigma_3)(1 + x\sigma_4\sigma_3)(1 + x\sigma_2'\sigma_4)$$

$$= 4\cosh^3 K(1 + x^3\sigma_1'\sigma_2') = e^{-g(K)} \cosh K'(1 + x'\sigma_1'\sigma_2') = e^{-g(K) + K'\sigma_1'\sigma_2'}.$$

$$(5.25)$$

The expression (5.25) has the form of the Boltzmann factor $\exp(K'\sigma_1'\sigma_2')$ with the renormalized constant $K' = \tanh^{-1}(\tanh^3 K)$ or $x' = x^3$—this formula and $g(K) = \ln(\cosh K'/4 \cosh^3 K)$ are called recursion relations. The partition function of the whole system in the new variables is $\sum_{\{\sigma'\}} \exp\left[-g(K)N/3 + K' \sum_i \sigma_i'\sigma_{i+1}'\right]$. The term proportional to $g(K)$ represents the contribution to the free energy of the short-scale degrees of freedom, which have been averaged out. This term does not affect the statistics of the remaining variables, which is determined by (5.24) with the renormalization of the constant, $K \to K'$. Let us discuss this renormalization. Since $K \propto 1/T$, then $T \to \infty$ corresponds to $x \to 0+$ and $T \to 0$ to $x \to 1-$. One is interested in the values that do not change under the RG, i.e., that represent a fixed point of this transformation. Both $x = 0$ and $x = 1$ are fixed points of the transformation $x \to x^3$. The first one corresponds to the flat distribution $\rho\{\sigma_i\} = \text{const}$ having equal probabilities for both signs if every spin is independent of other spins. Such a distribution has a zero mean and

---

3. The term initially meant putting to death every tenth soldier of a Roman regiment that ran from a battlefield.

corresponds to a disordered state. The second fixed point corresponds to the distribution that is a delta function peaked at either $\sigma_i = 1$ or $\sigma_i = -1$ for all $i$. It is an ordered state with nonzero magnetization $\langle \sigma_i \rangle$.

The first fixed point is stable and the second one is unstable: iterating the process for $0 < x < 1$, we see that $x$ approaches zero and effective temperature infinity. That means that large-scale degrees of freedom are described by the distribution with the effective temperature so high that the system is in a disordered (paramagnetic) state. Long-range order is impossible in one-dimensional systems with short-range interaction because any overturned spin breaks the correlation between left and right parts. For however small yet finite temperature, the distance to the next overturned spin is $e^K$, which is finite so that the system is disordered at larger distances. At this limit, we have $K, K' \to 0$ so that the contribution of the small-scale degrees of freedom is independent of the temperature: $g(K) \to -\ln 4$. We see that spatial rescaling leads to the renormalization of temperature: the spin chain looks hotter when viewed with less resolution.

What entropic measure monotonically changes along RG to quantify the irreversibility of forgetting? Eliminating some degrees of freedom decreases the entropy of the whole system even when RG moves us toward a more disordered state. Then it is more natural to be interested in the entropy per spin or in the mutual information between eliminated and remaining degrees of freedom. These two entropic measures are related. For RG, we can define the mutual information between two sublattices: eliminated and remaining. The positivity of the mutual information then implies the monotonic growth of the entropy per spin, $h(K) = \lim_{N \to \infty} S(K, N)/N$. Consider, for instance, the RG eliminating every second spin, $N \to N/2$, and renormalizing $\tanh K' = \tanh^2 K$ (such RG has the same flow and the same fixed points). Subtracting the entropy of the original lattice from the sum of the entropies of two identical sublattices gives the mutual information: $I = 2S(N/2, K') - S(N, K) = N[h(K') - h(K)] \geq 0$. That shows that in 1D the entropy per block spin grows with the block size upon RG at any distance from the fixed point.

Let us now consider a finite $N$ system that comes close to a fixed point. In a finite system with short-range correlations, the entropy for large $N$ is generally as follows:

$$S(N) = hN + C, \qquad I = N[h(K') - h(K)] + 2C' - C. \qquad (5.26)$$

We now have two characteristics, $h$ and $C$. In a fixed point, the extensive terms in $I$ cancel and $I = C > 0$. This is why $C$ is called *excess entropy*. One can explain the positivity of $C$ by saying that a finite system appears more random than it is since we haven't seen all the possible correlations.

Mutual information also naturally appears in the description of the information flows in the real space. Let us break the 1D $N$ chain into two parts, $M$ and $N - M$. The mutual information between two parts of the chain (or between the past and the future of a message) is as follows: $I(M, N - M) = S(M) + S(N - M) - S(N)$. Here the extensive parts (linear in $M, N$) cancel in the limit $N, M \to \infty$. Therefore, such mutual information is equal to $C$ from (5.26). Note that the past-future mutual information also serves as a measure of the message complexity (that is, the difficulty of predicting the message).

After these general arguments, let us now compute $h$ and $C$ for the Ising model. Recall that the entropy is expressed via the partition function as follows:

$$S = \frac{E - F}{T} = T\frac{\partial \ln Z}{\partial T} + \ln Z.$$

For the 1D Ising chain, $Z = 2(2 \cosh K)^{N-1}$ gives $h = \ln(2 \cosh K) - K \tanh K$ and $C = K \tanh K - \ln(\cosh K)$. Upon RG flow, these quantities monotonously change from $h(K) \approx 3e^{-2K}$, $C \approx \ln 2$ at $K \to \infty$ to $h(K) \approx \ln 2$, $C \to 0$ at $K \to 0$. One can interpret this by saying that $C = \ln q$, where $q$ is the degeneracy of the ground state. Indeed, $q = 2$ at the zero-temperature fixed point due to two ground states with opposite magnetization, while $q = 1$ in the fully disordered state. So this mutual information (and the excess entropy) measures how much information per one degree of freedom one needs to specify. (For noninteger $q$ obtained midway through the RG flow, one can think of it as viewing the system with finite resolution.)

The RG flow is rather trivial in 1D, where RG moves systems toward disorder so that $K' < K$ and $h(K') > h(K)$. In higher dimensions, there could exist fixed points (limiting distributions) that describe neither a low-temperature fully ordered state nor a high-temperature fully disordered state, but a critical state of the phase transition between the two. This is described in section A.9.

**Exercise 5.1:** RG and the family of universal distributions.
Consider a set of random iid variables $x_1 \ldots x_N$.

(a) The RG reduces the number of random variables by replacing any two of them by their mean (half sum):

$z_i = (x_{2i-1} + x_{2i})/2$. Show that the Fourier image of the distribution $\rho(k) = exp(-|k|)$ is a fixed point of this map. Study the linear stability of this fixed point. What probability density does this correspond to? Why doesn't this contradict the central limit theorem?

(b) Consider the one-parametric family of the transformations:
$z_i = (x_{2i-1} + x_{2i})/2^{1/\mu}$. Find the fixed point, that is, the distribution invariant under this transformation.

# 6

# Fundamental Limits
# of Uncertainty

In our consideration so far, an uncertainty could be arbitrarily small or large. Physical laws impose restrictions. First, the quantum nature of the world bounds uncertainty from below, making some information unavailable in principle. Second, gravity and relativity create regions of space (black holes) from where no information can escape; surprisingly, this imposes an upper bound on how much information can be stored in a finite region of space. That upper bound contains all three fundamental constants known to physics: the gravitational constant, the speed of light, and the Planck constant.

The first three sections of this chapter are devoted to the implications for information theory of the quantum nature of our world. That adds some unavoidable uncertainty, which cannot be diminished by improving measurement precision or gathering more information. Quantum description is in principle incomplete as it is formulated in physical terms that describe the possible results of measurements, which are interactions with a classical object that does not itself obey quantum laws. The unique source of quantum uncertainty is superposition: a quantum system can be in many different states at the same time. Measurement chooses one state, which irreversibly changes the system. This change cannot be made arbitrarily small. The results of quantum measurement are then truly random (not because we did not bother to learn more about the system). Accounting for quantum-mechanical uncertainty is thus of fundamental value for information theory.

Interest in quantum information is also pragmatic. Quantum superposition means that evolution proceeds in the space of factorially more dimensions than the classical system. This is a source of the parallelism of quantum computations. Moreover, classical systems, including computers, are limited by locality; that is, operations have only local effects. Spatially separated quantum systems may be entangled with each other so that operations may have nonlocal effects. Those two basic facts motivate an interest in quantum computations and communications, which is briefly discussed in the fourth and fifth sections.

The last section of this chapter is devoted to the upper bounds of uncertainty imposed by the existence of black holes, considered as gates out of the accessible world. We shall see how general relativity and quantum mechanics conspire to impose some fundamental restrictions on the amount of information in the world.

## 6.1   Quantum Mechanics and Entropy

Not surprisingly, quantum information theory is also based on the notion of entropy, which is similar to classical entropy yet differs in some important ways. Uncertainty and probability already exist in quantum mechanics, where we consider an isolated system. On top of that, we shall consider quantum statistics due to incomplete knowledge, which is caused by considering subsystems. Here I give a very brief introduction to the subject, focusing on information and entropy and their most dramatic differences from the classical world. Recall that the entropy consideration by Planck is what started quantum physics in the first place. Looking at two asymptotics of a spectral curve, he decided to search for an analytic formula matching their *entropies*, simply adding them. The resulting formula is the logarithm of the number of ways to distribute a given energy in equal discrete portions—quantization was born.

Quantum mechanics mathematically is quite elementary, since it is based on linear algebra, that is, the study of vectors and linear operations on them. A quantum state of a physical system is a vector, which contains all the information. We denote such (column) vectors either by $\psi_i$ or by the Dirac notation $|i\rangle$. The dual (row) vector then is denoted $\langle i|$ and the inner (scalar) product by $\langle i|j\rangle$. If in some orthonormal basis $\{|i\rangle\}$, two vectors are presented as $|v\rangle = \sum_i v_i |i\rangle$ and $|w\rangle = \sum_i w_i |i\rangle$, then $\langle v|w\rangle = \sum_i v_i^* w_i$. A property that can be measured is called an observable and is described by a self-adjoint operator (matrix), say, $\hat{O}$. The expectation value of an observable in a state $\psi$ is an inner product $\langle \psi| \hat{O} |\psi\rangle$.

The fundamental statement is that any system can be in a single state, $\psi_i$, or in a superposition of states, $\psi = \sum_i a_i \psi_i$, where $a_i$ are generally complex numbers. An example of a single state is a fixed-energy eigenstate of a Hamiltonian (which is an operator that is a matrix). The possibility of a superposition is the total breakdown from classical physics, where those states (say, with different energies) are mutually exclusive.

There are two things we can do with a quantum state: either let it evolve without touching or measure it. Measurement is classical; it produces one and only one state from the initial superposition, and immediately repeated measurements produce the same outcome. However, repeated measurements of the identically prepared initial superposition, $\psi = \sum_i a_i \psi_i$, find different states: the state $i$ appears with probability $p_i = |a_i|^2$.

There is already an uncertainty in any state of an isolated quantum system. A product of two operators $\hat{X}\hat{Z}$ defines the result of the successive measurements. If the two observables $X$ and $Z$ can simultaneously have definite values, then $\hat{Z}\hat{X}$ gives the same result as $\hat{X}\hat{Z}$. However, matrices generally do not commute. If the operators are noncommuting, $[\hat{X}, \hat{Z}] = \hat{X}\hat{Z} - \hat{Z}\hat{X} \neq 0$, then the observables cannot simultaneously have definite values since the product of their variances is restricted from below:

$$|\langle\psi|[\hat{X}, \hat{Z}]|\psi\rangle|^2 = 4|\langle\psi|\hat{X}\hat{Z}|\psi\rangle|^2 - |\langle\psi|\hat{X}\hat{Z} + \hat{Z}\hat{X}|\psi\rangle|^2$$

$$\leq 4|\langle\psi|\hat{X}\hat{Z}|\psi\rangle|^2 \leq 4\langle\psi|\hat{X}^2|\psi\rangle\langle\psi|\hat{Z}^2|\psi\rangle. \tag{6.1}$$

Here the second step is the Cauchy-Bunyakovsky-Schwarz inequality. In particular, momentum and coordinate are such a pair: $\hat{X} = \hat{\mathbf{p}} - \langle\mathbf{p}\rangle, \hat{Z} = \hat{\mathbf{q}} - \langle\mathbf{q}\rangle$. Since the momentum operator in the coordinate representation is $\hat{p}_x = \imath\hbar\partial_x$, then $[\hat{p}_x, x] = -\imath\hbar$, which gives the Heisenberg uncertainty principle: the variances of the coordinate and the momentum along the same direction, $\sigma_p = \langle p^2\rangle - \langle p\rangle^2, \sigma_q = \langle q^2\rangle - \langle q\rangle^2$, satisfy the inequality

$$\sqrt{\sigma_p\sigma_q} \geq \hbar/2. \tag{6.2}$$

That means that we cannot describe quantum states as points in the phase space $(p, q)$. What we call "quantum entanglement" is ultimately related to the fact that one cannot localize quantum states in a finite region of the phase space—if coordinates are fixed somewhere, then the momenta are not.

The variances depend on the state. One can show (see exercise 6.1) that a Gaussian wave packet corresponds to the minimal product of variances and turns (6.2) into equality. For the corresponding Gaussian distribution, the entropy $S_G$ is a logarithm of the variance. Taking the log of the Heisenberg equality for Gaussian states, we obtain $\log(2\sigma_p/\hbar) + \log(2\sigma_q/\hbar) = S_G(p) + S_G(q) = 0$, recasting it as a relation on the (differential) entropies of the Gaussian probability distributions of the momentum and the coordinate. It turns into an inequality for arbitrary distributions. In $d$ dimensions, different components commute, so that $\sqrt{\sigma_\mathbf{p}\sigma_\mathbf{q}} \geq d\hbar/2$ and

$$S(\mathbf{p}) + S(\mathbf{q}) \geq \log d. \tag{6.3}$$

More formally, if two matrices do not commute, they cannot be diagonalized by a single orthonormal set. Assume that two noncommuting matrices $\hat{X}$ and $\hat{Z}$ can be respectively diagonalized by (projected onto) two different orthonormal bases, $\{|x\rangle\}$, $\{|z\rangle\}$. If we measure a quantum state $\psi$ by projecting onto the $x$ basis, the outcomes define a classical probability distribution $p(x)$. The Shannon entropy $S(X)$ quantifies how uncertain we are about the outcome before we perform the measurement. There is also a corresponding classical probability distribution of outcomes when we measure the same state $\psi$ in the $z$ basis. The two bases are incompatible, so there is a trade-off between our uncertainty about $X$ and about $Z$, captured by the inequality

$$S(X) + S(Z) \geq \log(1/c), \quad c = \max_{x,z} |\langle x|z\rangle|^2. \tag{6.4}$$

We see that the lower bound on the total uncertainty is given by the maximum overlap between any two eigenvectors, that is, by the degree of mutual nonorthogonality of the two bases. We prove a more general form of that relation in the next subsection.

Two different bases, $\{|x\rangle\}$, $\{|z\rangle\}$, for a $d$-dimensional space are called mutually unbiased if $|\langle x_i|z_k\rangle|^2 = 1/d$ for all $i, k$. That means that if we measure any $x$-basis state in the $z$ basis, all $d$ outcomes are equally probable and give the same contribution to the total probability: $\sum_k |\langle x_i|z_k\rangle|^2 = \sum_i |\langle x_i|z_k\rangle|^2 = 1$. For measurements in two mutually unbiased bases performed on a pure state, the entropic uncertainty relation becomes

$$S(X) + S(Z) \geq \log d. \tag{6.5}$$

This inequality is saturated by $x$-basis states, for which $S(X) = 0$ and $S(Z) = \log d$. In one dimension, $\log d = 0$.

Note that the right-hand sides of (6.2, 6.3, 6.4, 6.5) are fixed lower bounds, in contrast to that of (6.1), which generally depends on the state $\psi$ and is thus not universal.

*Qubit*    So far, we have dealt with the statistics of the measurement outcomes, and the entropy has been the familiar classical Gibbs-Shannon entropy. Let us now deal with the *states* of quantum systems rather than with the measurements. We have defined a classical "bit" as a unit of information choosing between two states, so we can also call a bit a physical system, where we distinguish two states only. This could be a coin, a magnetic moment looking along or against an applied field, a photon with two polarizations, etc. Similarly, we define a *qubit* as a quantum system having only two orthogonal states: $|0\rangle$ and $|1\rangle$. The most general state of a qubit $A$ is a superposition of two states, $\psi_A = a\,|0\rangle + b\,|1\rangle$, where any observable is as follows:

$$\langle\psi_A|\hat{O}_A|\psi_A\rangle = |a|^2\langle0|\hat{O}_A|0\rangle + |b|^2\langle1|\hat{O}_A|1\rangle + (a^*b + ab^*)\langle0|\hat{O}_A|1\rangle. \tag{6.6}$$

Normalization requires $|a|^2 + |b|^2 = 1$, and if the overall phase does not matter, then a qubit is characterized by two real numbers—say, the amplitude $|a|$ and the relative phase between $a$ and $b$. Alternatively, we may characterize it by a complex number. The qubit represents the unit of quantum information the same way the bit represents the unit of classical information. Apparently, quantum systems operate with much more information—one needs many bits to record a complex number with reasonable precision, and the difference grows exponentially when we compare the states of $N$ classical bits with the possible states of $N$ qubits. Moreover, a qubit is not a classical bit because it can be in a superposition; nor can it be considered a random ensemble of classical bits with the probability $|a|^2$ in the state $|0\rangle$, because the phase difference of the complex numbers $a, b$ matters, as seen from (6.6).

And yet quantum mechanics tells us that we cannot measure the complex numbers $a, b$, that is, we cannot determine the quantum state of the qubit. This is in sharp contrast with our ability to determine the state of a bit (say, when a classical computer retrieves memory). Measurements of a qubit bring either the result $|0\rangle$ with the probability $|a|^2$ or the result $|1\rangle$ with the probability $|b|^2 = 1 - |a|^2$. In other words, a quantum coin can defy gravity and stand on its edge at an arbitrary angle, but any measurement collapses it on one side, either heads or tails up. What use then in quantifying the quantum information if we cannot measure it? One should not despair, though. While we cannot measure it directly, we can communicate it. Moreover, we describe below indirect ways to manipulate a quantum system so that a measurement gives a result, which depends distinctly on the state of the system. These ways involve entanglement between different subsystems.

***Exercise 6.1:*** Least uncertain wave packet.

Proceeding from the fact that the momentum operator in the coordinate representation is $\hat{p}_x = \imath \hbar \partial_x$, find the state $\psi(x)$ that minimizes the expectation of the product of variances of the coordinate and momentum. What is the corresponding $\psi(p)$?

## 6.2   Quantum Statistics and the Density Matrix

To consider subsystems, we need to pass from quantum mechanics to quantum statistics and introduce the fundamental notion of the *density matrix*. Consider a composite system $AB$, which is in a state $\psi_{AB}$. If states of $A$ are characterized by $N$ vectors and of $B$ by $M$ vectors, we need to characterize $AB$ by $MN$ vector. We can make such a vector by the so-called tensor product,[1] multiplying every component of one vector state of $A$ by every component of one of $B$: $\psi_{AB} = \psi_A \otimes \psi_B$. This corresponds to independent subsystems. In this case, any operator $\hat{O}_A$ acting only on $A$ has the expectation value ($\hat{I}$ is the identity operator)

$$\langle \psi_{AB}|\hat{O}_A \otimes \hat{I}_B|\psi_{AB}\rangle = \langle \psi_A|\hat{O}_A|\psi_A\rangle \langle \psi_B|\hat{I}_B|\psi_B\rangle = \langle \psi_A|\hat{O}_A|\psi_A\rangle,$$

so that one can forget about $B$ and characterize $A$ by the vector $\psi_A$, as expected for independent systems. However, a general state $\psi_{AB}$ is not a single (tensor) product of $A$ and $B$ states. For example, if $A$ and $B$ are qubits, then a general state of a two-qubit system is a superposition, $a|00\rangle + b|11\rangle + c|01\rangle + d|10\rangle$. When $ab \neq cd$, such a superposition cannot be presented as a single tensor product:

$$(\alpha |0\rangle_A + \beta |1\rangle_A) \otimes (\alpha' |0\rangle_B + \beta' |1\rangle_B)$$

$$= \alpha\alpha' |00\rangle + \beta\beta' |11\rangle + \alpha\beta' |01\rangle + \alpha'\beta |10\rangle.$$

For arbitrary orthonormal bases, $\phi_A^i$ and $\phi_B^j$, one may generally expect a double sum, $\psi_{AB} = \sum_{ij} \alpha_{ij}\phi_A^i \otimes \phi_B^j$. Fortunately, a so-called singular value decomposition allows one to represent the matrix of the coefficients as $\alpha_{ij} = u_{ik}d_{kk}v_{kj}$, where $\hat{u}, \hat{v}$ are unitary and $\hat{d}$ is diagonal. We then define $\psi_A^k = u_{ki}\phi_A^i$ and $\psi_B^k = v_{jk}\phi_B^j$. This is called Schmidt decomposition by orthonormal vectors $\psi_A^k, \psi_B^k$, which allows us to present any state of $AB$ as a single sum of the

---

1. For example, $(a, b) \otimes (c, d, e) = (ac, ad, ae, bc, bd, be)$.

products: for each vector from $A$, there is just one vector from $B$:

$$\psi_{AB} = \sum_k \sqrt{p_k}\psi_A^k \otimes \psi_B^k. \tag{6.7}$$

If there is more than one term in this sum, we call subsystems $A$ and $B$ *entangled*. There is no factorization of the dependencies in such a state. We can always make $\psi_{AB}$ a unit vector so that $\sum_i p_i = 1$, and these numbers can be treated as probabilities (to be in the state $i$). Now the operator acting only on $A$ has the following expectation value:

$$\langle \psi_{AB}|\hat{O}_A \otimes \hat{I}_B|\psi_{AB}\rangle = \sum_{i,j} \sqrt{p_i p_j}\langle \psi_A^i|\hat{O}_A|\psi_A^j\rangle \langle \psi_B^i|\hat{I}_B|\psi_B^j\rangle$$

$$= \sum_{i,j} \sqrt{p_i p_j}\langle \psi_A^i|\hat{O}_A|\psi_A^j\rangle \delta_{ij} = \sum_i p_i \langle \psi_A^i|\hat{O}_A|\psi_A^i\rangle = \mathrm{Tr}_A \rho_A \hat{O}_A,$$

where we define the density matrix as

$$\rho_A = \sum_i p_i|\psi_A^i\rangle \langle \psi_A^i|. \tag{6.8}$$

Tr denotes the trace, which is the sum of the diagonal elements of a (square) matrix. The density matrix is all we need to describe $A$. From now on, we shall distinguish pure states described by a vector and mixed states described by a density matrix. The matrix is Hermitian; it has all nonnegative eigenvalues and a unit trace. Every matrix with those properties can be "purified," that is, presented (nonuniquely) as a density matrix of the subsystem $A$ in the extended system $AB$, which as a whole is in a pure state, $\psi_{AB}$. The possibility of purifications is quantum mechanical with no classical analog: the classical analog of a density matrix is a probability distribution, which cannot be purified.

The statistical density matrix describes a mixed state or, in other words, an ensemble of states. Different ensembles can give the same density matrix; see exercise 6.2. A mixed state described by a matrix must be distinguished from a quantum-mechanical superposition described by a vector. The superposition is in both states simultaneously; the ensemble is in perhaps one or perhaps the other, characterized by probabilities—that uncertainty appears because we do not have any information on the state of the $B$ subsystem.

We characterize the uncertainty in classical physics by a probability vector $\{p_i\}$ and in quantum mechanics by a state vector $\psi_i$. In quantum statistics, we need a matrix, generally nondiagonal, whose $ij$ element quantifies how the states $i$ and $j$ of $A$ are correlated via all possible states of $B$.

***Example 6.1:*** Consider a pure entangled quantum state of a two-qubit system $A, B$:[2]

$$\psi_{AB} = a \,|00\rangle + b \,|11\rangle . \tag{6.9}$$

One can predict one qubit by knowing another. Any operator acting on $A$ gives

$$\langle\psi_{AB}|\hat{O}_A \otimes \hat{I}_B|\psi_{AB}\rangle = (a^*\langle 00| + b^*\langle 11|)\hat{O}_A \otimes \hat{I}_B|(a|00\rangle + b|11\rangle)$$

$$= |a|^2\langle 0|\hat{O}_A|0\rangle + |b|^2\langle 1|\hat{O}_A|1\rangle. \tag{6.10}$$

That corresponds to a diagonal density matrix:

$$\rho_A = \mathrm{Tr}_B\left(|a|^2\,|00\rangle\,\langle 00| + |b|^2\,|11\rangle\,\langle 11| + a^*b\,|00\rangle\right.$$

$$\langle 11| + ab^*\,|11\rangle\,\langle 00|\left.\right)$$

$$= |a|^2\,|0\rangle\,\langle 0| + |b|^2\,|1\rangle\,\langle 1| = \begin{bmatrix} |a|^2 & 0 \\ 0 & |b|^2 \end{bmatrix}. \tag{6.11}$$

We can interpret this as saying that the system $A$ is in a mixed state, that is, with probability $|a|^2$ in the quantum state $|0\rangle$ and with probability $|b|^2$ in the state $|1\rangle$. Due to the orthogonality of $B$ states, the same results $(6.10, 6.11)$ are obtained if $\langle 0|\hat{O}_A|1\rangle \neq 0$ and for whatever relative phase between $a$ and $b$, in contrast to $(6.6)$. Being in a superposition is not the same as being in a mixed state, where the relative phases of the states $|0\rangle, |1\rangle$ are experimentally inaccessible.

In 1852, long before Landau and von Neumann introduced the quantum density matrix (1927), Stokes used an equivalent description for a partially polarized light. Let us consider the (electric) field in a propagating wave: $\mathbf{E} = \mathbf{A}(t)\exp(\iota\mathbf{k}\cdot\mathbf{r} - \iota\omega t)$. The two-dimensional polarization vector $\mathbf{A}$ is perpendicular to $\mathbf{k}$. Since the wave has both amplitude and phase, it is described by a complex vector, like a quantum state. Time dependence $\mathbf{A}(t)$ means that polarization (slowly) changes with time. If one measures the light intensities (i.e., the quadratic functions of the field), the only nonzero averages over time windows exceeding $1/\omega$ are $J_{ab} = \overline{E_a E_b^*}$. We then can characterize the polarization by the Hermitian $2 \times 2$ density matrix with a unit trace: $\rho_{ab} = J_{ab}/\mathrm{Tr}\hat{J}$. Here uncertainty appears due to finite temporal resolution.

---

2. An early idea of entanglement was conjured up in the seventeenth century: it was claimed that if two magnetic needles were magnetized at the same place and time, they would stay "in sympathy" forever at however large distances, and the motion of one would be reflected on the other. One con man tried to sell this communication device to Galileo, who didn't buy it.

**Exercise 6.2:** Density matrix.

Consider two mixed states (ensembles): In the first ensemble $A$, the system can be in the state $|0\rangle$ with the probability $3/4$ and in the state $|1\rangle$ with the probability $1/4$. In the second ensemble $B$, the system can be in the state $|a\rangle = \sqrt{3/4}\,|0\rangle + \sqrt{1/4}\,|1\rangle$ and in the state $|b\rangle = \sqrt{3/4}\,|1\rangle - \sqrt{1/4}\,|0\rangle$ with equal probability.

(a) Write the density matrices for these two ensembles in the basis $|0\rangle, |1\rangle$.

(b) Consider two sets of normalized vectors, $|\psi_i\rangle$ and $|\phi_j\rangle$, and two probability distributions, $p_i$ and $q_j$. The sets are related by $\sqrt{p_i}\,|\psi_i\rangle = \sum_j u_{ij}\sqrt{q_j}\,|\phi_j\rangle$, where the matrix is unitary: $\sum_i u_{ij}u_{ik}^* = \delta_{jk}$. Find the relation between two density matrices, $\rho_1 = \sum_i p_i\,|\psi_i\rangle\langle\psi_i|$ and $\rho_2 = \sum_j q_j\,|\phi_j\rangle\langle\phi_j|$.

## 6.3   Entanglement Entropy

One can ascribe to any density matrix $\rho_A$ the entropy by the formula analogous to the Gibbs-Shannon entropy of a probability distribution (von Neumann 1927, 1932):

$$S(\rho_A) = -\mathrm{Tr}\,\rho_A \log \rho_A. \tag{6.12}$$

Since we are dealing with diagonalizable matrices, a logarithm (or any other function) of the matrix is defined for a diagonal matrix: if $\rho = \sum_k p_k\,|k\rangle\langle k|$, then $\log \rho = \sum_k \log(p_k)\,|k\rangle\langle k|$. To avoid confusion, we always use Greek letters for the argument of von Neumann entropy and Latin letters for the argument of Shannon entropy.

The von Neumann entropy quantifies the type of uncertainty, which exists only in a quantum world and is related to the principal restriction of measurements to a finite volume. The classical entropy is the logarithm of the number of microstates compatible with the given macroscopic state. The quantum entropy $S(\rho_A)$ is, roughly speaking, the logarithm of the number of states of the inaccessible part $B$ of the universe compatible with all measurements of $A$, together with a priori knowledge that $A + B$ is in a pure state.

Evidently, $S(\rho_A)$ is invariant under a unitary transformation, $\rho_A \to U\rho_A U^{-1}$, which is an analog of the Liouville theorem on the conservation of distribution by Hamiltonian evolution. Just like the classical entropy, it is nonnegative, equals to zero only for a pure state, and reaches its maximum $\log d$ for equipartition (when all $d$ nonzero eigenvalues are equal), that is, it satisfies concavity (2.7). What does not have a classical analog is that the purifying system $B$ has the same entropy as $A$ (since the same $p_i$ appears in its density matrix). Moreover, von Neumann entropy of a part $S(\rho_A)$ can be larger than that of the whole system $S(\rho_{AB})$. When $AB$ is pure, $S(\rho_{AB}) = 0$, but $S(\rho_A)$ could be nonzero (Landau 1927). Information can be encoded in the correlations among the parts, yet be invisible when we look at one part of a quantum system. That purely quantum correlation between different parts is called entanglement, and the von Neumann entropy of a subsystem of pure state is called *entanglement entropy*.

Classically, we measure the nonlocality of information encoding by the mutual information $I(A, B) = S(A) + S(B) - S(A, B)$, which never exceeds the sum of two entropies. Quantum $I$ is nonnegative like classical, but generally is different. The nonlocality of information encoding is raised to a whole new level in the quantum world. For example, when $AB$ is in an entangled pure state, then $S(\rho_{AB}) = 0$ so that $A$ and $B$ together are perfectly correlated, but separately each one is in a mixed state with $S(\rho_A) = S(\rho_B) > 0$. Classically, the mutual information of perfectly correlated quantities is equal to each of their entropies, but quantum mutual information is their sum that is twice more: $I(\rho_{AB}) = S(\rho_A) + S(\rho_B) - S(\rho_{AB}) = 2S(\rho_A)$. Quantum correlations are stronger than classical.

The von Neumann entropy of a density matrix is the Shannon entropy $S(p) = -\sum_i p_i \log p_i$ of its vector of eigenvalues, which is the probability distribution $\{p_i\}$ of its orthonormal eigenstates. In particular, for $\psi_{AB} = a \, |00\rangle + b \, |11\rangle$, we have $S(\rho_A) = -|a|^2 \log_2 |a|^2 - |b|^2 \log_2 |b|^2$. The maximum $S(\rho_A) = 1$ is reached when $|a|^2 = |b|^2 = 1/2$, which is called a state of maximal entanglement. In this case, when we trace out $B$ (or $A$), we wipe out the information about the whole: any measurement on $A$ or $B$ cannot tell us anything about the state of the pair since both outcomes are equally probable. On the contrary, when either $b \to 0$ or $a \to 1$, the entropy $S(\rho_A)$ goes to zero, and measurements (of either $A$ or $B$) give us definite information on the state of the pair.

The original von Neumann argument involved a mixing process. Consider a gas of molecules, where $pN$ are in a pure state $|a\rangle$ and $(1 - p)N$ are in an orthogonal state $|b\rangle$. This is described by the power $N$ of the density matrix $\rho = p \, |a\rangle \langle a| + (1 - p) \, |b\rangle \langle b|$. Orthogonality of the states means that $a$ molecules can be separated from $b$ molecules, for instance, by a wall permeable only for one state. We then double our volume and move $a$, $b$ walls from opposite ends to the center to separate the molecules completely. The densities in the respective halves are different: their ratio is $p/(1-p)$. We now squeeze one half by the factor $p$ and the other by $(1-p)$, making the densities equal and returning the whole volume to the original value. We can now transform $a$ states into $b$ states by some unitary transformation, say, by temporal evolution. Here the quantum nature shows up. After that, we remove the partition and obtain a zero-entropy state. Since the squeezing at a constant temperature $T$ releases the heat $-T[p\log p + (1-p)\log(1-p)] = TS(p)$, which decreases the entropy by $S(p)$, it is the mixing entropy of the original mixture.

Let $\rho = \sum_k p_k |\psi^k\rangle \langle \psi^k|$ be diagonal in the basis of eigenvectors $\{|\psi^k\rangle\}$, but we measure by projecting $\rho$ on a different orthogonal set $\{|\phi^i\rangle\}$. In this case, the outcome $i$ happens with the probability $q_i = \langle \phi^i|\rho|\phi^i\rangle = \sum_k p_k D_{ik}$, where $D_{ik} = |\langle \phi^i|\psi^k\rangle|^2$ is a so-called double stochastic matrix, that is, $\sum_i D_{ik} = \sum_k D_{ik} = 1$. The Shannon entropy of that probability distribution is larger than the von Neumann entropy,

$$S(q) = S(p) + \sum_{ik} p_k D_{ik} \log \left( \sum_n p_n D_{in}/p_k \right)$$
$$= S(p) + D(q|p) \geq S(p) = S(\rho),$$

that is, such measurements are less predictable. Mathematically, the interpretation is that the diagonal elements $(q_i)$ are more random than the eigenvalues $(p_k)$ for a nonnegative Hermitian matrix.

*General uncertainty relation*   If we measure a *mixed* state $\rho$ by projecting onto the orthonormal basis $\{|x\rangle\}$, the outcomes define the density matrix $\hat{M}_x \rho = \rho_x = \sum_x |x\rangle \langle x|\rho|x\rangle \langle x|$. The measurement operator $\hat{M}_z$ projecting onto another basis $\{|z\rangle\}$ defines $\hat{M}_z \rho = \rho_z = \sum_z |z\rangle \langle z|\rho|z\rangle \langle z|$. Both density matrices are diagonal so that each von Neumann entropy is equal to the corresponding Shannon entropy: $S(\rho_x) = S(X)$ and $S(\rho_z) = S(Z)$. We now introduce the relative entropy for density matrices:

$$D(\rho|\rho_x) = \text{Tr}\,\rho\,(\log \rho - \log \rho_x) = \text{Tr}\,\rho \log \rho$$
$$- \text{Tr}\,\rho_x \log \rho_x = S(X) - S(\rho).$$

Here we use the property of trace: $\text{Tr}\,\rho \log \rho_x = \text{Tr}\,\rho_x \log \rho_x$. As in the classical case, $D$ is nonnegative and quantifies the number of measurements needed to distinguish two density matrices. It also possesses the important property of monotonicity, that is, nonincreases upon any partial trace. This property is intuitive, as in the classical case—after all, it should be no easier to distinguish two density matrices looking only at the subsystem, yet the proof is complicated and we do not give it here. We now use monotonicity of the relative entropy $D(\rho|\rho_x)$ under the action of the measurement in the $z$ basis:

$$D(\rho|\rho_x) \geq D(\hat{M}_z\rho|\hat{M}_z\rho_x) = D(\rho_z|\hat{M}_z\rho_x) = -S(Z) - \text{Tr } \rho_z \log \hat{M}_z\rho_x.$$
(6.13)

The new density matrix obtained by two measurements,

$$\hat{M}_z\rho_x = \hat{M}_z\hat{M}_x\rho = \sum_z |z\rangle \sum_x \langle z|x\rangle \langle x|\rho|x\rangle \langle x|z\rangle \langle z|,$$

is diagonal, so that

$$\log \hat{M}_z\rho_x = \sum_z |z\rangle \log \left(\sum_x \langle z|x\rangle \langle x|\rho|x\rangle \langle x|z\rangle\right) \langle z|.$$

The logarithm is a monotonic function:

$$\log \left(\sum_x \langle z|x\rangle \langle x|\rho|x\rangle \langle x|z\rangle\right) \leq \log \left(\max_{x,z} |\langle x|z\rangle|^2 \sum_x \langle x|\rho|x\rangle\right)$$

$$= \log \left(\max_{x,z} |\langle x|z\rangle|^2\right).$$

Substituting that into (6.13), we obtain the generalization of the uncertainty relation for a mixed state:

$$S(X) + S(Z) \geq \log(1/c) + S(\rho), \quad c = \max_{x,z} |\langle x|z\rangle|^2.$$

Both sources of uncertainty in quantum statistics are here: nonorthogonality of states quantified by $c$ and entanglement quantified by $S(\rho)$. Comparing that with the uncertainty relations (6.4, 6.5) written for a pure state, we see that the von Neumann entropy quantifies the increase in uncertainty due to entanglement with the environment.

*Coming to equilibrium*   When a classical system is attached to a thermostat, it comes to thermal equilibrium with it, attaining entropy maximum determined by the temperature of the thermostat. But what if a quantum system is attached to a large system with which they together form a pure quantum state with zero entropy? Are thermalization and entropy growth possible for subsystems of a quantum system that as a whole remains in a pure quantum state? Yes, they are! Thermalization takes place for any subsystem of a large system if the dynamics are ergodic and can be characterized by the growth of the entanglement entropy. Then the system as a whole acts as a thermal reservoir for its subsystems, provided they are small enough.

Consider a small quantum system that at some moment is attached to a large system. At this moment, the information is encoded locally, the entanglement entropy is zero, and the subsystem is not in equilibrium with the whole system. As the small subsystem starts interacting with the large system and approaches equilibrium, the von Neumann entropy grows and reaches its maximum. Information, which was initially encoded locally in an out-of-equilibrium state, becomes encoded more and more nonlocally as the system evolves, eventually becoming invisible to an observer confined to the subsystem. Such thermalization can be quantified by a relative entropy. Denote the (evolving) density matrix of our subsystem as $\rho$. If the evolution of the subsystem, when it is closed, is described by the Hamiltonian $H$, we can define the Gibbs density matrix as

$$\rho_0 = \frac{\exp(-\beta H)}{\operatorname{Tr}\,\exp(-\beta H)} = Z^{-1}\exp(-\beta H). \tag{6.14}$$

We now define the respective free energies via the energy and the von Neumann entropy:

$$F(\rho) = E - \beta^{-1}S(\rho) = \operatorname{Tr}\rho H + \beta^{-1}\operatorname{Tr}\rho\ln\rho = \beta^{-1}\operatorname{Tr}\rho(\ln\rho + \beta H),$$

$$F_0 = \beta^{-1}\operatorname{Tr}\rho_0(\ln\rho_0 + \beta H) = -\beta^{-1}\ln Z = -\beta^{-1}\ln\operatorname{Tr}\,\exp(-\beta H).$$

The relative von Neumann entropy between $\rho$ and $\rho_0$ can be expressed via the difference in the free energies:

$$D(\rho|\rho_0) = \operatorname{Tr}\rho\ln\rho - \operatorname{Tr}\rho\ln\rho_0$$

$$= \operatorname{Tr}\rho(\ln\rho + \beta H) + \beta^{-1}\ln\operatorname{Tr}\,\exp(-\beta H) = \beta[F(\rho) - F_0] \geq 0,$$

$$\tag{6.15}$$

where the last inequality follows from the positivity of the relative entropy, just like in the classical case (2.30). Therefore, the Gibbs state has the lowest free energy at a given temperature, which is determined by its environment treated as a thermostat. Unitary evolution of the subsystem and its environment induces on a subsystem a decrease (by monotonicity) of $D(\rho, \rho_0)$, eventually bringing the subsystem to the Gibbs state.

*Exercise 6.3:* Von Neumann entropy.

Consider two nonorthogonal states, $|0\rangle$ and the superposition $|s\rangle = (|0\rangle - |1\rangle)/\sqrt{2}$, mixed with the respective probabilities $p$ and $1 - p$. Find the density matrix $\rho$ in the orthogonal basis $|0\rangle$, $|1\rangle$, diagonalize it, compute the von Neumann entropy $S(\rho)$, and compare it with $S(p)$.

## 6.4   Quantum Communications

Without going into the specifics of quantum processors and communication schemes, here we discuss how much information one can transfer by sending (or sharing) quantum objects. Let us first ask, How many bits of classical information can be recovered from a quantum system? Even though any qubit potentially contains a complex number, any measurement only gives one or another state, so that a pure state of a qubit can store one classical bit. The four orthogonal maximally entangled states of the qubit pair, $(|00\rangle \pm |11\rangle)/\sqrt{2}$ and $(|01\rangle \pm |10\rangle)/\sqrt{2}$, can store two bits. Generally, when sending a quantum system whose state is determined by a $d$-dimensional complex vector, one can send at most $\log d$ bits of classical information (for instance, by sending one of the states from $d$ basic vectors).

How this is related to the quantum mutual information can be realized by looking at a more symmetric variant of the same problem. Let a composite system $AB$ be described by the density matrix $\rho_{AB}$. Alice has access to $A$, while Bob has access to $B$. The results of the measurements belong to classical information and can be written in the notebooks $C_A$ and $C_B$. The maximal number of bits Alice can get from her measurements about those of Bob (and vice versa) is the classical mutual information between their notebooks, $I(C_A, C_B)$. Measurements correspond to tracing out some degrees of freedom so that monotonicity guarantees that $I(C_A, C_B) \leq I(\rho_{AB}) \leq \log d$, where $I(\rho_{AB}) = S(\rho_A) + S(\rho_B) - S(\rho_{AB})$ is the mutual information of the initial density matrix $\rho_{AB}$.

Let us now turn to quantum information, that is, to the information about quantum states themselves rather than to the measurement results. We pose the same natural question we asked for classical communications in section 2.2: How much can a message be compressed? That is, what is the maximum information one can transmit per quantum state? Is it given by von Neumann entropy or by Shannon entropy, as in the classical case? Now the letters of our message are quantum states picked with their respective probabilities $p_k$; that is, each letter is described by the density matrix and the message is a tensor product. Leaving aside how actual quantum communication devices handle information compression, we discuss here only the amount of quantum information, that is, the number of combinations of states involved. If the states are mutually orthogonal and the density matrix is diagonal, it is essentially the classical case; that is, the answer is given by the Shannon entropy $S(p) = -\sum_k p_k \log p_k$, which is the same as the von Neumann entropy in this case. For example, the output of a qubit source that sends $|0\rangle$ with probability $p$ and $|1\rangle$ with probability $1 - p$ can be compressed similarly to the classical source described in section 2.2.

The new issue in quantum theory is that nonorthogonal states cannot be perfectly distinguished, a feature with no classical analog. If a pure state $AB$ is built from nonorthogonal states taken with the weights $p_i$, then the density matrix $\rho_A$ is nondiagonal. There is then the difference between the Shannon entropy of the mixture and the von Neumann entropy of the matrix, $S\{p_i\} - S(\rho_A)$. It is nonnegative and quantifies how much distinguishability is lost when we mix nonorthogonal pure states. Measuring $\rho_A$, we receive $S(\rho_A)$ bits, which is less than $S\{p_i\}$ bits that were encoded mixing the states with probabilities $\{p_i\}$.

For example, nonorthogonal states $|0\rangle$ and the superposition $|s\rangle = (|0\rangle + |1\rangle)/\sqrt{2}$ cannot be distinguished when a measurement in the basis $|0\rangle$, $|1\rangle$ brings $|0\rangle$. Consider an output producing $|0\rangle$ with the probability $p$, and $|s\rangle$ with the probability $1 - p$, like in exercise 6.3. Sending classical information about these states brings the information $S(p)$. Could we use different encoding corresponding to a shorter mean length of a codeword? The letters of our alphabet, $|0\rangle$ and $|s\rangle$, both contain the state $|0\rangle$, which means redundancy. The redundancy must allow for tighter compression than $S(p)$. That can be demonstrated using essentially the argument from section 2.2 with the only difference being that, instead of typical sequences, we consider typical subspaces. Could we decrease the entropy using the orthogonal states $|0\rangle$, $|1\rangle$? A long $N$ string emitted by the source looks like a superposition of the terms having, after reordering, the following form:

$$|0\rangle^{\otimes Np} |s\rangle^{\otimes N(1-p)} \approx |0\rangle^{\otimes N(1+p)/2} |1\rangle^{\otimes N(1-p)/2}. \tag{6.16}$$

This is because in the limit $N(1-p) \gg 1$ the product $|s\rangle^{\otimes N(1-p)}$ can be approximated by the superposition of the states with equal probability of $|0\rangle$ and $|1\rangle$. The number of the states of the form (6.16) is given by the number of $N(1+p)/2$ choices out of $N$; the logarithm of the number of states is $NS(1/2+p/2)$ according to the Stirling formula. If $(1+p)/2 > 1-p$, that is, $p > 1/3$, we can use the states of the form (6.16) as the new alphabet letters and neglect atypical states. We then achieve compression for sending classical information about these states, since $S(p) - S(1/2+p/2) = (1/2+p/2)\log(1/2+p/2) + (1/2-p/2)\log(1/2-p/2) - p\log p - (1-p)\log(1-p) > 0$ for $p > 1/3$. That bound makes sense, since at $p < 1/3$ the chosen way of encoding actually increases redundancy, and one needs to use a different encoding.

The most efficient encoding uses the states, where the density matrix of our source is diagonal, instead of the states $|0\rangle$, $|s\rangle$ or $|0\rangle$, $|1\rangle$, where it is not. The orthonormal eigenvectors can be found in the solution to exercise 6.3 in the appendix. The number of typical strings is then given by the Shannon entropy of this representation, which is now equal to the von Neumann entropy and is strictly lower than the Shannon entropy for any representation where the density matrix is not diagonal.

The best rate of quantum information transfer (the number of qubits carried per letter of a long message) is given by the von Neumann entropy of the density matrix of the source only when we deal with a mixture of pure states. This is not true when $\rho = \sum_k p_k \rho_k$ and $\rho_k$ are mixed states. It is easy to see from a trivial example: Suppose that a particular mixed state $\rho_0$ with $S(\rho_0) > 0$ is chosen with probability $p_0 = 1$. Then the message $\rho_0 \otimes \rho_0 \otimes \ldots$ carries no information.

When our alphabet is made of mixed yet mutually orthogonal states, the states are distinguishable and the problem is classical, since we can just send the probabilities of the states; so the maximal rate is the Shannon entropy $S(p)$. However, it is *less than* the von Neumann entropy, which now includes a nonzero entropy of every mixed state $\rho_k$. Because all $\rho_k$ are orthogonal, they could be made diagonal simultaneously, and we obtain

$$S(\rho) = -\sum_k \text{Tr}(p_k \rho_k) \log(p_k \rho_k)$$

$$= -\sum_k \left( p_k \log p_k + p_k \text{Tr}\, \rho_k \log \rho_k \right) = S(p) + \sum_k p_k S(\rho_k).$$

This shows that, when $\rho_k$ are mixed states, $S(\rho)$ is no longer a good measure of quantum entanglement since it clearly mixes quantum and classical correlations. In this case, von Neumann entropy exceeds Shannon entropy:

$$S(p) = S(\rho) - \sum_k p_k S(\rho_k) = S\left(\sum_k p_k \rho_k\right) - \sum_k p_k S(\rho_k). \qquad (6.17)$$

To conclude, the information transfer rate

i) by orthogonal pure states is equal to $S(p) = S(\rho)$,
ii) by nonorthogonal pure states is equal to $S(\rho)$, which is less than $S(p)$,
iii) by orthogonal mixed states is equal to $S(p)$, which is less than $S(\rho)$.

For nonorthogonal mixed states, it is believed that

$$\chi(\rho_k, p_k) = S\left(\sum_k p_k \rho_k\right) - \sum_k p_k S(\rho_k)$$

(called in quantum communications Holevo information) defines the limiting compression rate in all cases, including when it does not coincide with $S(p)$. The reason for this belief is that $\chi$ is monotonic (i.e., it decreases when we take partial traces), but $S(\rho)$ is not—one can increase von Neumann entropy by going from a pure to a mixed state. It follows from concavity that $\chi$ is always nonnegative. We see that it depends on the probabilities $p_k$, that, is on the way we prepare the states. Of course, (6.17) is a kind of mutual information; it tells us how much, on average, the von Neumann entropy of an ensemble is reduced when we know which preparation was chosen. It is exactly like classical mutual information, $I(A, B) = S(A) - S(A|B)$, which tells us how much the Shannon entropy of $A$ is reduced once we get the value of $B$. So we see that classical Shannon information is mutual von Neumann information. One also calls $\chi$ the accessible information of an ensemble of quantum states, that is, the maximal number of bits of information that can be acquired about the preparation of the state on average.

## 6.5   Conditional Entropy and Teleportation

Similar to the classical conditional entropy (2.12), one defines for von Neumann entropy

$$S(\rho_{AB}|\rho_B) = S(\rho_{AB}) - S(\rho_B). \qquad (6.18)$$

However, this is not an entropy conditional on something known; moreover, it is not zero for correlated quantities but negative! Indeed, for pure $AB$, one has $S(\rho_{AB}|\rho_B) = -S(\rho_B) < 0$. Classical conditional entropy measures how many classical bits we need to add to $B$ to fully determine $A$. Similarly, we would expect quantum conditional entropy to measure how many qubits Alice needs to send to Bob to reveal herself. But what does it mean when $S(\rho_{AB}|\rho_B)$ is negative? In this situation, it includes the amount of quantum information that Bob already shares with Alice.

That negativity is due to entanglement between $A$ and $B$, which allows the trick of *teleportation*. Teleportation moves quantum states around without a quantum channel, and we shall see below that negative von Neumann conditional entropy counts the number of possible future teleportations. Imagine that Alice has in her possession a qubit $A_0$ and wants Bob to create in his lab a qubit in a state identical to $A_0$. However, she is only able to communicate by sending a classical message. If Alice knows the state of her qubit, there is no problem (except that communicating a complex number exactly requires an infinite number of classical bits): she tells Bob (say, over the telephone) the state of her qubit and he creates one like it in his lab. If, however, Alice does not know the state of her qubit, all she can do is make a measurement, which will give some information about the prior state of qubit $A_0$. She can tell Bob what she learns, but the measurement will destroy the remaining information about $A_0$ and it will never be possible for Bob to re-create it. So she needs to make a measurement revealing no information about $A_0$. Then what information can that measurement reveal? It must be about something else that Alice and Bob share.

Suppose then that Alice and Bob have previously shared a qubit pair $A_1, B_1$ in a known entangled state, for example,

$$\psi_{A_1B_1} = \frac{1}{\sqrt{2}}(|00\rangle + |11\rangle)_{A_1B_1}. \tag{6.19}$$

Bob then took $B_1$ with him, leaving $A_1$ with Alice. In this case, Alice can solve the problem by making a joint measurement of her system $A_0A_1$ in a basis, which is chosen so that no matter what the answer is, Alice learns nothing about the prior state of $A_0$. In that case, she also loses no information about $A_0$. But after getting her measurement outcome, she knows the full state of the system and she can tell Bob what to do to re-create $A_0$. To see how this works, let us describe a specific measurement that Alice can make on $A_0A_1$ that will shed no light on the state of $A_0$. The measurement must be a projection on a state where the probability of $A_0$ to be in the state $|0\rangle$ is exactly equal to the probability to be in the state $|1\rangle$. The following four states of $A_0A_1$ satisfy that property:

$$\frac{1}{\sqrt{2}}(|00\rangle \pm |11\rangle)_{A_0A_1}, \quad \frac{1}{\sqrt{2}}(|01\rangle \pm |10\rangle)_{A_0A_1}. \tag{6.20}$$

The states are chosen to be entangled, that is, having $A_0$ and $A_1$ correlated. We don't use the state with $|00\rangle \pm |10\rangle$, which has equal probability of zero and one for $A$ but no correlation between the values of $A_0$ and $A_1$.

Denote the unknown initial state of the qubit $A_0$ as $\alpha |0\rangle + \beta |1\rangle$; then the initial state of $A_0A_1B_1$ is

$$\frac{1}{\sqrt{2}}(\alpha |000\rangle + \alpha |011\rangle + \beta |100\rangle + \beta |111\rangle)_{A_0A_1B_1}. \tag{6.21}$$

Let's say that Alice's measurement, that is, the projection on the states (6.20), reveals that $A_0A_1$ is in the state

$$\frac{1}{\sqrt{2}}(|00\rangle - |11\rangle)_{A_0A_1}. \tag{6.22}$$

That means that only the first and the last terms in (6.21) contribute (with equal weights but opposite signs). After that measurement, $B_1$ will be in the state $(\alpha |0\rangle - \beta |1\rangle)_{B_1}$, whatever the (unknown) values of $\alpha, \beta$. Appreciate the weirdness of the fact that $B_1$ was uncorrelated with $A_0$ initially, but instantaneously acquired correlation after Alice performed her measurement a thousand miles away. Knowing the state of $B_1$, Alice can send two bits of classical information, telling Bob that he can re-create the initial state $\alpha |0\rangle + \beta |1\rangle$ of $A_0$ by multiplying the vector of his qubit $B_1$ by the matrix $\begin{bmatrix} 1 & 0 \\ 0 & -1 \end{bmatrix}$, which switches the sign of the second vector of the basis. The beauty of it is that Alice learned and communicated not what the state $A_0$ was, but how to re-create it.

To understand the role of the quantum conditional entropy (6.18) in teleportation, we symmetrize and purify our problem. Generally, the weirdness of quantum entropies can be traced to the purely quantum possibility of purification. Notice that $A_1$ and $B_1$ are maximally entangled (come with the same weights), so that $S(\rho_B) = \log_2 2 = 1$. On the other hand, $A_1B_1$ is in a pure state, so its von Neumann entropy is zero. Let us now add another system $R$, which is maximally entangled with $A_0$ in a pure state $A_0R$, say,

$$\psi_{A_0R} = \frac{1}{\sqrt{2}}(|00\rangle + |11\rangle)_{A_0R}. \tag{6.23}$$

Neither Alice nor Bob have access to $R$. From this viewpoint, the combined system $RAB = RA_0A_1B_1$ starts in a pure state, which is a direct product $\psi_{RA_0} \otimes \psi_{A_1B_1}$. Since $A_0$ is maximally entangled with $R$, then $S(\rho_{A_0}) = \log_2 2 = 1$, which is equal to the entropy of the $AB$ system, $S(\rho_{A_0A_1B_1}) = S(\rho_{A_0}) = 1$, since $A_1B_1$ is a pure state. Therefore, $S(\rho_{AB}|\rho_B) = S(\rho_{A_0A_1}|\rho_B) = S(\rho_{A_0A_1B_1}) - S(\rho_{B_1}) = 0$. One can show that teleportation is only possible when $S(\rho_{AB}|\rho_B)$ is nonpositive.

Recall that, classically, $S(A|B)$ measures how many bits of information Alice has to send to Bob (in addition to $B$, which he has) so that he will have full knowledge of $A$. The quantum analog of this involves qubits rather than classical bits. Suppose that $S(\rho_{AB}|\rho_B) > 0$ and Alice nevertheless wants Bob to re-create her states. She can simply send her states. The alternative is to do teleportation, which requires sharing with Bob an entangled pair for every qubit of her state to be teleported. Either way, Alice must be capable of *quantum communication*, that is, of sending a quantum system while maintaining its quantum state. For teleportation, she first creates some maximally entangled qubit pairs and sends half of each pair to Bob. Each time she sends Bob half of a pair, $S(\rho_{AB})$ is unchanged but $S(\rho_B)$ goes up by 1, so $S(\rho_{AB}|\rho_B) = S(\rho_{AB}) - S(\rho_B)$ goes down by 1. So $S(\rho_{AB}|\rho_B)$, if positive, is the number of such qubits that Alice must send to Bob to make $S(A|B)$ nonpositive and to make teleportation possible without any further quantum communication. Negative quantum conditional entropy measures the number of possible future qubit teleportations. We thus see that entanglement is an important resource in quantum communications.

## 6.6  The Way Out Is via a Black Hole

All things physical are information-theoretic in origin.

—JOHN WHEELER, 1990

A black hole presents a way to eliminate all uncertainty about a system by swallowing and forever eliminating it from our reach. No body, no uncertainty. On the other hand, this information, while inaccessible, still remains a part of our world. Our religious belief that uncertainty in the world can only increase leads us to the entropy of a black hole and to the ultimate restriction on the amount of information that can be encoded in a physical system.

*Area law*    Relativity imposes a restriction: no material body can move faster than light. Coupled with gravity, it makes possible inescapable regions of space—black holes. A body of the mass $m$ can escape a gravitating body of the mass $M$ from the distance $R$ if the kinetic energy $mv^2/2$ exceeds the potential energy $GMm/R$, where $G$ is the gravitational constant. Since $v < c$, then $R > r_h = 2GM/c^2$. One cannot escape from within the so-called horizon $r_h$, since the speed needed for that exceeds $c$. A black hole is an object, whose size is smaller than its horizon.

Since the interior is inaccessible, one may think that the entropy of a black hole must be zero for us. Let us now add to the mix the quantum theory that does not allow strict separation of space regions. The quantum entanglement entropy (between interior and exterior) is thought to be responsible for the entropy of black holes. To estimate it, we need an equation of state, that is, the relation between energy and temperature. The energy of the hole is simply $E_{BH} = Mc^2 = c^4 r_h/2G$. The temperature of the hole is determined by its radiation, which is due to a purely quantum phenomenon of particle-antiparticle pairs appearing from vacuum fluctuations. Such pairs usually stay together and soon annihilate. If, however, such a pair straddles the horizon, then the inside part is absorbed by the hole, while the outside part can escape and be registered as radiation (this is how the entanglement appears). The typical wavelength of such radiation can be estimated as $r_h$, and its energy/temperature is then $T \simeq \hbar c/r_h$. Now we can obtain the entropy (up to $\pi$ and order-unity factors) by integrating the equation of state $T = dE/dS$:

$$T \simeq \frac{\hbar c}{r_h} \simeq \frac{dE_{BH}}{dS_{BH}} \simeq \frac{c^4}{G}\frac{dr_h}{dS_{BH}} \Rightarrow S_{BH} \simeq \frac{r_h^2 c^3}{G\hbar}.$$

The entropy is proportional to the squared horizon, that is, to area rather than volume. Since any entropy is dimensionless, $\hbar G/c^3$ must be a square of some fundamental length. It is called the Planck length, $l_p = \sqrt{\hbar G/c^3} \simeq 10^{-17}$ cm, and it is the only combination with that dimensionality of the three fundamental physical constants, $c$, $\hbar$, $G$; it is the scale where quantization of gravity is expected to be important.[3] The entanglement entropy of a black hole can

3. One-parameter theories: $G$-theory, seventeenth century; $c$-theory, nineteenth and twentieth centuries; $\hbar$-theory, early twentieth century. Two-parameter theories: $c$, $G$ (general relativity), $c$, $\hbar$ (quantum electrodynamics), twentieth century. Hopefully, $c$, $G$, $h$-theory will appear in the twenty-first century.

thus be written as follows:

$$S_{BH} \simeq \frac{r_h^2 c^3}{G\hbar} = \frac{r_h^2}{l_p^2} \simeq \frac{GM^2}{\hbar c}. \tag{6.24}$$

The area law behavior of the entanglement entropy in microscopic theories could be related to the holographic principle—the conjecture that the information contained in a volume of space can be completely encoded by the degrees of freedom that live on the boundary of that region.

*Bekenstein bound*    We can now estimate the information capacity (not a channel capacity!) defined as the maximal amount of information that can be encoded in a system by exploiting all of its degrees of freedom down to the quantum level. Is there a universal limit on how large the entropy of a physical system can be? The answer is given by the so-called Bekenstein bound (and its generalizations). On dimensional grounds, it can be guessed as follows. The entropy must be the total energy $E$ (including any rest masses) divided by a temperature (in energy units). The temperature must be determined by the system size $R$—the smaller the size, the higher the temperature. Confining a system to a smaller region by quantum uncertainty increases the kinetic energy. The only combination with the dimensionality of energy one can make out of $R$ and the world constants $\hbar$, $c$ is the same $\hbar c/R$, which is the energy of a photon with wavelength $R$. That suggests a bound in the following form: $S \leq RE/\hbar c$ (Bekenstein 1981, 2004; Casini 2008).

That bound was argued by exploiting the only known way to eliminate entropy from the observable world—to drop it into a black hole. If we drop a body of energy $E$ and entropy $S$ into a black hole of large mass $M \gg E/c^2$, then the black hole's mass will grow by $E/c^2$. According to (6.24), the entropy of the hole will then grow by $\simeq 2GME/\hbar c^3$ plus a negligible term of order $E^2$. Meanwhile, the entropy $S$ has gone forever out of this world. The second law then requires that $S < 2GME/\hbar c^3 = r_h E/\hbar c$. This is expected to hold up to a body size comparable with the black hole horizon, which gives the estimate for the bound:

$$S \leq \frac{RE}{\hbar c}. \tag{6.25}$$

We presume that the body itself is not a black hole, that is, its size exceeds its horizon, $R > r_h(E) \simeq GE/c^4$. Substituting $E < c^4 R/G$ into (6.25), we write the entropy restriction solely in terms of the radius:

$$S \leq \frac{R^2 c^3}{G\hbar} = \frac{R^2}{l_p^2}.$$

(6.26)

Comparing (6.26) and (6.24), we conclude that a system must be a black hole to realize the capacity limit. We note without elaboration that (6.25, 6.26) actually refer to the difference between the entropy of the system with energy $E$ and the entropy of the quantum vacuum in the region of size $R$.

In the thermodynamic limit, the classical total entropy is extensive; that is, it is proportional to the system volume or total number of degrees of freedom. We now see that the entropy is proportional to the volume only as long as one can squeeze more and more distinguishable matter into it. When there is so much matter or so little space that the system turns into a black hole, we can see only the horizon, and the entropy is proportional to the area (like a hologram where a 3D image is encoded on a 2D surface).

The appearance of the gravitational constant $G$ in (6.26) deserves reflection. Via black holes, gravity makes some information unavailable in principle, which is a source of the bound. Another way to look at it is that black holes provide the gates out of the observable world. A counterpart to this is the Big Bang, which provided a gate into this world—how something comes out of nothing could probably teach us important lessons about the nature of information as well. It is also worth bearing in mind that gravitational instabilities soon after the Big Bang are the origin of structures in the universe and thus of most of information.

# 7

# Conclusion

This chapter attempts to compress the book to its most essential elements.

## 7.1 Take-Home Lessons

1. Thermodynamics studies restrictions imposed by the hidden on the observable. It deals with two extensive quantities. The first one (energy), $E$, is conserved for a closed system, and its changes are divided into work (due to observable degrees of freedom) and heat (due to hidden ones). The second quantity (entropy), $S$, can only increase for a closed system and reach its maximum in thermal equilibrium, where the system entropy is a convex function of the energy. All available states lie below this convex curve in the $S, E$ plane.

2. Convexity of the dependence, $E(S)$, allows us to introduce temperature as the derivative of the energy with respect to the entropy. The extremum of the entropy means that the temperatures of the connected subsystems are equal in equilibrium. The same is true for the energy derivatives with respect to volume and other extensive variables. The entropy increase (called the second law of thermodynamics) imposes restrictions on thermal engine efficiency, that is, the fraction of heat used for work:

$$\frac{W}{Q_1} = \frac{Q_1 - Q_2}{Q_1} = 1 - \frac{T_2 \Delta S_2}{T_1 \Delta S_1} \leq 1 - \frac{T_2}{T_1}.$$

If information processing generates $\Delta S = S_2 - S_1 = (Q - W)/T_2 - Q/T_1$, its energy price is as follows:

$$Q = \frac{T_2 \Delta S + W}{1 - T_2/T_1}.$$

3. Need in statistics appears due to incomplete knowledge: we are able to follow only part of the degrees of freedom and only with finite precision. Statistical physics defines the (Boltzmann) entropy of a closed system as the log of the phase volume, $S = \log \Gamma$, and assumes (for lack of knowledge) the uniform distribution $w = 1/\Gamma$, called microcanonical. For a subsystem, the (Gibbs) entropy is defined as the mean phase volume: $S = - \sum_i w_i \log w_i$. The probability distribution is then obtained, requiring maximal entropy for a given mean energy: $\log w_i \propto -E_i$. Information theory generalizes this approach; see point 11 below.

4. We quantify the lack of knowledge by the amount of information needed to make the knowledge complete and remove the uncertainty. We start by receiving information as answers to yes-no questions (called bits). The amount of information is the number of such answers, that is, $\log_2 n$, where $n$ is the number of possibilities each with the probability $1/n$ (Boltzmann entropy). If the probabilities $p_i$ are different from $1/n$, then the Shannon-Gibbs entropy/information is the mean logarithm: $- \sum_i p_i \log_2 p_i$. Convexity of the function $-p \log p$ guarantees that the information/entropy has its maximum for equal probabilities (when our ignorance is maximal).

5. A simple mathematical notion of convexity is a powerful tool. We first use it in thermodynamics to make sure that the extremum is on the boundary of the region and to make Legendre transforms of thermodynamic potentials. Concavity of the logarithm and entropic measures (including relative and von Neumann entropies) play a central role in classical and quantum statistics. Convexity is used to establish hierarchies and find the extremum. Convexity of the exponential function is used to show that, even when the mean of a random quantity is zero, its mean exponent exceeds unity. That provides for an exponential separation of trajectories in an

incompressible flow and exponential growth of the density of an element in a compressible flow. On the other hand, if the mean exponent is unity, $\langle e^{-\Delta S} \rangle = 1$, then the mean itself is negative: $-\langle \Delta S \rangle \leq 0$.

6. In our discrete thinking, we use another basic mathematical object—the sum of independent random numbers, $X = \sum_{i=1}^{N} y_i$. Three concentric statements can be made (see section A.2). The weakest one is that $X$ approaches its mean value, $\bar{X} = N \langle y \rangle$, exponentially fast in $N$. The next statement is that the distribution $P(X)$ is Gaussian in the vicinity of the width $\simeq N^{-1/2}$ around the maximum. The whole distribution is also very sharp, which is described by the large-deviation form: $P(X) \propto e^{-NH(X/N)}$, where $H \geq 0$ and $H(\langle y \rangle) = 0$. Applying this to the log of the probability of a given sequence, $\lim_{N \to \infty} \log p(y_1 \ldots y_N) = -NS(Y)$, we learn two lessons: i) the probability is independent of a sequence for most of them (almost all events are almost equally probable), ii) the number of typical sequences *grows exponentially and the entropy is the rate.*

7. The number of typical binary sequences of length $N$ is then $2^{NS}$, which cannot exceed $2^N$. The efficient encoding of the typical sequences thus involves words with lengths from unity to $NS$, which is less than $N$ if the probabilities of 0 and 1 are not equal. That means that the entropy is both the mean and the fastest rate of the reception of information brought by long messages/measurements. To squeeze out all the unnecessary bits, encoding is used both in industry and in nature.

8. If the transmission channel $B \to A$ makes errors, then the message does not completely eliminate uncertainty; what remains is the conditional entropy, $S(B|A) = S(A, B) - S(A)$, which is the mean rate of growth of the number of possible errors. Sending extra bits to correct these errors lowers the transmission rate from $S(B)$ to the mutual information, $I(A, B) = S(B) - S(B|A) = S(A) + S(B) - S(A, B)$, which is the mean difference of the uncertainties before and after the message. The great news is that one can still achieve an asymptotically error-free transmission if the transmission rate is lower than $I$. The maximum of $I$ over all source statistics is the channel capacity, which is the maximal rate of asymptotically error-free transmission. In particular, to maximize the capacity of sensory processing, one makes the signal probability flat using the

response function of a sensor equal to a cumulative probability of stimuli.

9. Very often our goal is not to transmit as much information as possible but to compress it and process it as little as possible, looking for an encoding with a minimum of the mutual information. For example, rate-distortion theory looks for the minimal rate $I$ of information transfer under the restriction that the signal distortion does not exceed the threshold $\mathcal{D}$. This is done by minimizing the functional $I + \beta \mathcal{D}$.

10. Conditional probability allows for hypothesis testing by the Bayes rule: $P(h|e) = P(h)P(e|h)/P(e)$. That is, the probability $P(h|e)$ that the hypothesis $h$ is correct after we receive the data $e$ is the prior probability $P(h)$ times the support $P(e|h)/P(e)$ that $e$ provide for $h$. Taking a log and averaging, we obtain the familiar $S(h|e) = S(h) - I(e, h)$. The Bayes approach demonstrates that there is no inference without prior assumption. If our hypothesis concerns the probability distribution itself, then the difference between the true distribution $p$ and the hypothetical distribution $q$ is measured by the relative entropy $D(p|q) = \langle \log_2 (p/q) \rangle$. This is yet another rate—how the error probability grows with the number of trials. $D$ also measures the decrease of the transmission rate due to nonoptimal encoding: the mean length of the codeword is not $S(p)$ but is bounded by $S(p) + D(p|q)$. Mutual information is a particular case of relative entropy; they are both invariant with respect to arbitrary transformations of variables in a continuous case, which facilitates their ever-widening area of applications.

11. Since so much hangs on getting the right distribution, how best can we guess it from the data? This is achieved by maximizing the entropy under the given data—"the truth and nothing but the truth." That explains and makes universal the entropy maximization from point 3. What was thought to be a unique property of thermal equilibrium is now understood as universally applicable common sense. It also sheds new light on physics, telling us that, on some basic level, all states are constrained equilibria. Whenever we encounter a trade-off, free energy appears, whose two terms quantify the opposite tendencies. Not only do its (conditional) minima describe physical systems, but they are presently the most powerful technical tools of optimization, from our Bayesian brain to machine learning algorithms.

12. Information is physical. At a finite temperature, both learning and erasing information require work. The energetic price of a learning-erasing cycle is $T$ times the mutual information between the system and the measuring device. Another side of the physical nature of information is that there is a (Bekenstein) limit on how much entropy one can squeeze inside a given radius; surprisingly, the limit is proportional to the area rather than the volume and is realized by black holes—our gates out of this world.

13. Full knowledge persists, partial knowledge dissipates. Irreversible entropy growth may seem to contradict the laws of mechanics, which are time-reversible and preserve the $N$-particle phase-space density. If we follow precisely all the degrees of freedom, the entropy is conserved and no information is lost. But if we follow only part of them, the entropy of that part generally grows as it interacts with the rest—whatever information we had is getting less relevant with time. We illustrate that for a one-particle momentum distribution of a dilute gas. Assuming that before every collision particles are independent, one obtains the Boltzmann kinetic equation, which, in particular, describes the irreversible growth of the one-particle entropy. Therefore, the difference must grow between the growing sum of one-particle entropies and the constant total entropy. That difference describes correlations and is called mutual information. Similarly, the thermalization of a quantum subsystem increases the entanglement entropy since the information is getting encoded in interaction with the environment and is locally inaccessible.

14. Total entropy growth can appear even if we follow all the degrees of freedom but do it with finite precision, that is, if we consider the evolution of finite phase-space regions. Instability leads to the separation of trajectories, which spread over the whole phase space under generic reversible Hamiltonian dynamics, very much like flows of an incompressible liquid are mixing (metaphorically, extra digits in precision add new degrees of freedom for unstable systems). Spreading and mixing in phase space correspond to the approach to equilibrium and entropy growth. On the contrary, to deviate a system from equilibrium, one adds external forcing and dissipation, which makes its phase flow compressible and distribution nonuniform.

15. The renormalization group (RG) is the best known way to forget information. As always with forgetting, the trick is to choose what to

keep, which is decided by the renormalization. For example, we can divide the sum of two random numbers either by 2, keeping the mean, or by $\sqrt{2}$, keeping the variance. That leads to different asymptotic distributions, which is the main focus of RG. We find that the entropy of the partially averaged and renormalized distribution is the proper measure of forgetting in simple cases, like adding random numbers on the way to the central limit theorem. In physical systems with many degrees of freedom, the quantity that changes monotonically upon RG could be the mutual information, defined in two ways: either between remaining and eliminated degrees of freedom or between different parts of the same system.

16. Two central themes of quantum information and the two respective sources of quantum uncertainty are nonorthogonality and entanglement. The first theme appears in quantum mechanics, where uncertainty can be characterized by classical entropy. Quantum statistics appears when we treat subsystems and must deal with von Neumann entanglement entropy. The quantum entropy of the whole can be less than the entropy of a part. In particular, the whole system can be in a pure state with zero entropy, in which case all the entropy of a subsystem comes from entanglement.

17. The last lesson is two progressively more powerful forms of the second law of thermodynamics, which originally was $\langle \Delta S \rangle \geq 0$. The first new form, $\langle e^{-\Delta S} \rangle = 1$, is the analog of a Liouville theorem. The second form relates the probabilities of forward and backward processes: $\rho^{\dagger}(-\Delta S) = \rho(\Delta S)e^{-\Delta S}$.

## 7.2   Epilogue

The central idea of this book is that learning about the world means building a model, which is essentially finding an efficient representation of the data. Optimizing information transmission or encoding may seem like a technical problem, but it is actually the most important task of science, engineering, and survival. Science works on more and more compact encoding of the strings of data, which culminates in formulating a law of nature, potentially describing an infinity of phenomena.

The main mathematical tool we learned here is an ensemble equivalence in the thermodynamic limit; its analog is the use of typical sequences in communication theory. The result is two roles of entropy: it defines maximum transmission and minimum compression.

Another central idea is that entropy is not a property of the physical world, but the information we lack about it. And yet the information is physical—it has an energetic value and a monetary price. The difference between work and heat is that we have information about the former but not the latter. That means that one can turn information into work and one needs to release heat to erase information. We also have learned that one not only pays for information but can turn information into money as well. The physical nature of information is manifested in the universal limit on how much of it we can squeeze into a space restricted by a given area.

The panoramic view accepted here works on different levels. Natural scientists and engineers tend to see analogies between phenomena. One analogy extensively exploited here is that measurements, predictions, recording retrievals, etc. can all be treated and described uniformly as different forms of communication. Another analogy is between finding optimal strategy in economics (proportional gambling), biology (phenotype switching), engineering design, data processing, perceptual inference, etc. On a higher level, mathematicians see analogies between analogies. For the two analogies above, the unifying mathematical notions are relative entropy and free energy. Convexity with its bag full of inequalities is another recurring mathematical notion unifying different approaches to the classes of phenomena, rather than phenomena themselves.

No rigorous proofs were given in this book, replaced instead by examples or hand-waving arguments of varying plausibility. A more rigorous and detailed while still compact deductive presentation of thermodynamics can be found in *Thermodynamics* by Callen (1965). Detailed information theory with proofs can be found in *Elements of Information Theory* by Cowen and Thomas, whose chapter 1 gives a concise overview. A more practical and problem-oriented approach with numerous exercises can be found in *Information Theory, Inference and Learning Algorithms* by MacKay. Those interested in proofs for chapter 5 can find them in *An Introduction to Chaos in Nonequilibrium Statistical Mechanics* by Dorfman. On quantum information, the two comprehensive books are those by Preskill (2015), and Nielsen and Chuang (2010).

I also wish to stress that the examples given in this book represent a small slice of the ever-widening avalanche of applications; more biological applications can be found in *Biophysics* by Bialek, and others in original articles and reviews. Numerous references scattered through the text, like (Zipf 1949), give you the most compact encoding for a search. On the other end of the spectrum is the popular book *The Origins of Life: From the Birth of Life to the Origin of Language* by Maynard Smith and Szathmary, describing evolution as a set of transitions from competition (between replicating molecules, genes, cells, individuals) to cooperation, which raises competition to the level of collectives (chromosomes, multicellular organisms, animal groups, societies). Accompanied by the appearance and development of signal systems, from cellular to human languages, came major transitions in information storage and transfer. We mention the speculative but compelling hypothesis that the explosion of innovations that started approximately 50,000 years ago was brought about by the development of language, which broke barriers between social, technical, and foraging skills. This is another argument in favor of a panoramic view.

Several important subjects were left out of this book. Our focus was largely (though not entirely) on finding a data description that is good on average. There exists a closely related approach that focuses on finding the shortest description and ultimate data compression for a given string of data. The Kolmogorov complexity is defined as the shortest binary computer program able to compute the string. It allows us to quantify how much order and randomness is in a given sequence—a truly random sequence cannot be described by an algorithm shorter than itself, while any order allows for compression. Complexity is (approximately) equal to the entropy if the string is drawn from a random distribution, but is actually a more general concept, treated in texts on computer science. A fundamental issue is the dramatic difference between the classical and quantum classifications of computational complexity. After much hesitation, I also left out evolutionary game theory, which describes, in particular, the appearance of cooperation and diversity—subjects having direct bearing on the information transfer in our world. Other subjects for mathematically oriented readers to explore are the applications of entropy as a measure of irreversibility in geometry (see, e.g., Perelman 2002) and the relation of entropy subadditivity to the rich world of isoperimetric and related inequalities (see, e.g., Gromov 2013).

Taking a wider view, I invite you to reflect on the history of our attempts to realize the limits of the possible, from heat engines to communication channels to computations. Will the next step be to study the natural limits of thinking and feeling?

Looking back, one may wonder why accepting the natural language of information took so much time and was so difficult for scientists and engineers. Generations of students (myself included) were tortured by "paradoxes" in statistical physics, which disappear when information language is used. I suspect that the resistance was to a large extent caused by the misplaced desire to keep the scientist out of science. A dogma that science must be something "objective" and only related to the things independent of our interest in them obscures the simple fact that science is a form of human language. True, we expect it to be objectively independent of the personality of this or that scientist as opposed, say, to literature, where we celebrate the difference between languages (and worlds) of Shakespeare and Tolstoy. However, science is the language designed by and for humans, so it necessarily reflects both the way body and mind operate and the restrictions on our ability to obtain and process data. Presumably, omnipresent and omniscient beings would have no need for the statistical information approach described here. As far as physics is concerned, I do not share the belief, widely held inside and outside the discipline, that physicists' notions are truly objective and fundamental, as opposed even to chemistry (where the distinction between organic and inorganic is due to our distinctively human interest in life), not to mention linguistics or economics. I believe that we, physicists, can benefit from better appreciating the essential presence of the scientist in science (for instance, to understand the special status of measurement in quantum mechanics).

As we learned here, better understanding must lead to a more compact presentation; hopefully, the next edition will be shorter.

# Extras, Exercises, and Solutions

This appendix addresses some advanced subjects, which are either more technical or more cross-disciplinary, or both.

## A.1 Formal Structure of Thermodynamics

Both energy and entropy are homogeneous, first-order functions of their variables: $S(\lambda E, \lambda V, \lambda N) = \lambda S(E, V, N)$ and $E(\lambda S, \lambda V, \lambda N) = \lambda E(S, V, N)$ (here $V$ and $N$ stand for the whole set of extensive macroscopic parameters). Differentiating the second identity with respect to $\lambda$ and taking it at $\lambda = 1$, one gets the Euler equation:

$$E = TS - PV + \mu N. \qquad (A.1)$$

The equations of state are homogeneous of zero order, for instance,

$$T(\lambda E, \lambda V, \lambda N) = T(E, V, N).$$

That confirms that the temperature, pressure, and chemical potential are the same for a portion of an equilibrium system as for the whole system.

Generally, thermodynamics can be developed for as many quantities as we observe. But what is the minimal number of observables for a meaningful description? It may seem that a thermodynamic description of a one-component mechanical system requires operating functions of three intensive variables. Let us show that the homogeneity leaves only two independent parameters. For example, the chemical potential $\mu$ can be found as a function of $T$ and $P$. By differentiating (A.1) and comparing with (1.5), one gets the so-called Gibbs-Duhem relation (in the energy representation), $N d\mu = -S dT + V dP$, or for quantities per mole, $s = S/N$ and $v = V/N$: $d\mu = -s dT + v dP$. In other words, one can choose $\lambda = 1/N$ and use first-order homogeneity to get rid of the variable $N$, for instance: $E(S, V, N) = NE(s, v, 1) = Ne(s, v)$. In the entropy representation,

$$S = E\frac{1}{T} + V\frac{P}{T} - N\frac{\mu}{T},$$

the Gibbs-Duhem relation again states that, because $dS = (dE + PdV - \mu dN)/T$, the sum of products of the extensive parameters and the differentials of the corresponding intensive parameters vanish:

$$Ed(1/T) + Vd(P/T) - Nd(\mu/T) = 0. \qquad (A.2)$$

Let us summarize the formal structure: The fundamental relation is equivalent to the three equations of state (1.4). If only two equations of state are given, then the Gibbs-Duhem relation may be integrated to obtain the third relation up to an integration constant; alternatively, one may integrate the molar relation $de = Tds - Pdv$ to get $e(s, v)$, again with an undetermined constant of integration.

**Example A.1:** Consider an ideal monatomic gas characterized by two equations of state (found, say, experimentally with $R \simeq 8.3 \, \text{J/mole K} \simeq 2 \, \text{cal/mole K}$):

$$PV = NRT, \qquad E = 3NRT/2. \qquad (A.3)$$

The extensive parameters here are $E, V, N$, so we want to find the fundamental equation in the entropy representation, $S(E, V, N)$. We write (A.1) in the form

$$S = E\frac{1}{T} + V\frac{P}{T} - N\frac{\mu}{T}. \qquad (A.4)$$

Here we need to express intensive variables $1/T, P/T, \mu/T$ via extensive variables. The equations of state (A.3) give us two of them:

$$\frac{P}{T} = \frac{NR}{V} = \frac{R}{v}, \qquad \frac{1}{T} = \frac{3NR}{2E} = \frac{3R}{2e}. \qquad (A.5)$$

Now we need to find $\mu/T$ as a function of $e, v$ using the Gibbs-Duhem relation in the entropy representation (A.2). Using the expression of intensive via extensive variables in the equations of state (A.5), we compute $d(1/T) = -3Rde/2e^2$ and $d(P/T) = -Rdv/v^2$, and substitute into (A.2):

$$d\left(\frac{\mu}{T}\right) = -\frac{3}{2}\frac{R}{e}de - \frac{R}{v}dv, \quad \frac{\mu}{T} = C - \frac{3R}{2}\ln e - R\ln v,$$

$$s = \frac{1}{T}e + \frac{P}{T}v - \frac{\mu}{T} = s_0 + \frac{3R}{2}\ln\frac{e}{e_0} + R\ln\frac{v}{v_0}. \qquad (A.6)$$

Here we assume that the system has the entropy $s_0$ in the state with the parameters $e_0, v_0$.

## A.2   Central Limit Theorem and Large Deviations

The true logic of this world is to be found in the theory of probability.

—JAMES MAXWELL

A bridge from statistical physics to information theory is a simple techni-cal tool used in both. Mathematics, underlying most of the statistical physics in the thermodynamic limit, comes from universality, which appears upon adding independent random numbers. The weakest statement is the law of large numbers: the sum approaches the mean value exponentially fast. The next level is the central limit theorem, which states that the majority of fluctuations around the mean have a Gaussian probability distribution. Consideration of large, rare fluctuations requires the so-called large-deviation theory. Here we briefly present all three at the physical (not mathematical) level.

Consider the variable $X$, which is a sum of many independent, iden-tically distributed (iid) random numbers, $X = \sum_1^N y_i$. Its mean value, $\langle X \rangle = N\langle y \rangle$ grows linearly with $N$. Here we show that its fluctuations $X - \langle X \rangle$ not exceeding $\mathcal{O}(N^{1/2})$ are governed by the central limit theorem: $(X - \langle X \rangle)/N^{1/2}$ becomes for large $N$ a Gaussian random variable with vari-ance $\langle y^2 \rangle - \langle y \rangle^2 \equiv \Delta$. The quantities $y_i$ that we sum can have quite arbitrary statistics $\mathcal{P}(y)$; the only requirements are that the first two moments, the mean $\langle y \rangle$ and the variance $\Delta$, are finite. Finally, the fluctuations $X - \langle X \rangle$ on the larger scale $\mathcal{O}(N)$ are governed by the large-deviation theorem, which states that there exists a convex function $H$ such that the probability distribution of $X$ asymptotically has the form

$$P(X) \propto e^{-NH(X/N)}. \qquad (A.7)$$

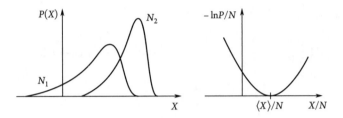

To show this, we write

$$P(X) = \int \delta \left( \sum_{i=1}^{N} y_i - X \right) P(y_1) dy_1 \ldots P(y_N) \, dy_N$$

$$= \int_{-\infty}^{\infty} dp \int \exp \left[ \imath p \left( \sum_{i=1}^{N} y_i - X \right) \right] P(y_1) dy_1 \ldots P(y_N) \, dy_N$$

$$= \int_{-\infty}^{\infty} dp e^{-\imath pX} \prod_{i=1}^{N} \int e^{\imath p y_i} P(y_i) dy_i = \int_{-\infty}^{\infty} dp e^{-\imath pX + NG(\imath p)}. \quad (A.8)$$

Here we introduce the generating function $\langle e^{zy} \rangle \equiv e^{G(z)}$. The derivatives of the generating function with respect to $z$ at zero are equal to the moments of $y$, while the derivatives of its logarithm $G(z)$ are called cumulants (see exercise A.3).

For large $N$, the integral (A.8) is dominated by the saddle point $z_0$ such that $G'(z_0) = X/N$. This is similar to representing the sum (1.10) by its largest term. If there are several saddle points, the result is dominated by the one with the largest probability. We assume that the contour of integration can be deformed in the complex plane $z$ to pass through the saddle point without hitting any singularity of $G(z)$. We now substitute $X = NG'(z_0)$ into $-zX + NG(z)$ and obtain the large-deviation relation (A.7) with

$$H = -G(z_0) + z_0 G'(z_0). \quad (A.9)$$

We see that $-H$ and $G$ are related by the ubiquitous Legendre transform (which always appear in the saddle-point approximation of the Fourier or Laplace transformations). Note that $NdH/dX = z_0(X)$ and

$$N^2 d^2 H/dX^2 = Ndz_0/dX = 1/G''(z_0).$$

The function $H$ of the variable $X/N - \langle y \rangle$ is called the Cramér or rate function since it measures the rate of probability decay with the growth of $N$ for every $X/N$. It is also sometimes called an entropy function since it is a logarithm of probability.

Several important properties of $H$ can be established independently of $G(z)$. It is a convex function as long as $G(z)$ is a convex function since their second derivatives have the same sign. It is straightforward to see that the logarithm of the generating function has a positive second derivative (at least for real $z$):

$$G''(z) = \frac{d^2}{dz^2} \ln \int e^{zy} P(y)\, dy$$

$$= \frac{\int y^2 e^{zy} P(y)\, dy \int e^{zy} P(y)\, dy - \left[\int y e^{zy} P(y)\, dy\right]^2}{\left[\int e^{zy} P(y)\, dy\right]^2} \geq 0. \quad \text{(A.10)}$$

This uses the Cauchy-Bunyakovsky-Schwarz inequality, which is a generalization of $\langle y^2 \rangle \geq \langle y \rangle^2$. Also, $H(z_0)$ takes its minimum at $z_0 = 0$, i.e., for $X$ taking its mean value $\langle X \rangle = N\langle y \rangle = NG'(0)$. The maximum of probability does not necessarily coincide with the mean value, but they approach each other when $N$ grows and the maximum gets very sharp—this is called the law of large numbers. Since $G(0) = 0$, the minimal value of $H$ is zero, that is, the probability maximum saturates to a finite value when $N \to \infty$. Any smooth function is quadratic around its minimum with $H''(0) = \Delta^{-1}$, where $\Delta = G''(0)$ is the variance of $y$. Quadratic entropy, $H \propto (X - \langle X \rangle)^2$, means Gaussian probability near the maximum—this statement is (loosely speaking) the essence of the central limit theorem. In the particular case of the Gaussian $P(y)$, the distribution $P(X)$ is Gaussian for any $X$. Non-Gaussianity of the $y$'s leads to a nonquadratic behavior of $H$ when deviations of $X/N$ from the mean are large, of the order of $\Delta/G'''(0)$.

One can generalize the central limit theorem and the large-deviation approach in two directions: i) for nonidentical variables $y_i$, as long as all variances are finite and none dominates the limit $N \to \infty$, it still works with the mean and the variance of $X$ being given by the average of means and variances of $y_i$; ii) if $y_i$ is correlated with a finite number of neighboring variables, one can group such "correlated sums" into new variables, which can be considered independent.

The above figure and (A.7, A.9) show how distribution changes upon adding more numbers. Is there any distribution that does not change upon averaging, that is, upon passing from $y_i$ to $\sum_{i=1}^{N} y_i/N$? That can be achieved for $H \equiv 0$, that is, for $G(z) = kz$, which corresponds to the Cauchy distribution $P(y) \propto (y^2 + k^2)^{-1}$. Since the averaging decreases the variance, it is no surprise that the invariant distribution has infinite variance. Distributions invariant under summation of variables are treated by the renormalization group in section 5.5.

*Exercise A.1:* Large deviations for the energy of particles.

Find the probability distribution of the kinetic energy, $E = \sum_1^N p_i^2/2$, of $N$ classical, identical unit-mass particles in 1D, which have the Maxwell distribution over momenta. Derive the large-deviation form of the distribution in the limit $N \to \infty$.

*Exercise A.2:* Large deviations for the binomial distribution.

One of the most widely used statistical distributions (including in this book) is the binomial distribution of two possible outcomes, $y = 1$ with probability $p$ and $y = 0$ with probability $1 - p$. Compute the probability that in a large number of trials $N$ the first outcome happens $qN$ times; that is, $X = \sum_{i=1}^N y_i = qN$. Do it in two ways: 1) discrete combinatoric, using the binomial formula $C_N^{qN} = N!/(qN)!(N - qN)!$ for the number of ways to choose $qN$ out of $N$, and the Stirling formula $\ln N! \approx N \ln N$; 2) continuous, using the large-deviation theory, that is, computing the cumulant generating function $G(z) = \ln\langle e^{zy} \rangle = \ln(pe^z + 1 - p)$ and the Legendre transform of it.

*Exercise A.3:* Generating function for cumulants.

The derivatives at zero of the logarithm of the generating function $G(z) = \ln\langle e^{zy} \rangle$ are called cumulants. Are $\kappa_n = (d^n G/dz^n)_{z=0}$ equal to the moments of $(y - \langle y \rangle)^n$? Express the first four $\kappa_1, \ldots, \kappa_4$ via $\mu_n = \langle y^n \rangle$.

## A.3  Numbers, Words, and Animal Signals

How best to encode numbers? Using a separate symbol for every number stops making sense when the number $N$ gets large. An alternative is to repeat one symbol $N$ times, but it is immediately clear that one can encode much more efficiently by dividing into groups so that the number $N$ can be encoded by $\log N$ symbols. One particular way of organizing numbers is the positional numeral system—a discovery of historical importance. Apart from the $\log N$ economy, another profound consequence of encoding where the meaning of the symbol depends on the position is that the decoding implies algebraic operations of multiplying and adding: $2024 = 2 \times 1000 + 2 \times 10 + 4$. Such a system allows simple, automatic rules for computations. This was formulated by the Persian mathematician al-Khwarizmi, from whose name the word *algorithm* is derived; *algebra, alcohol,* etc. also have Arabic origins.

Human languages encode meaning not in separate letters but in words. The oldest system of writing was logographic, where every word required a separate symbol, or logogram. Several of these systems were developed independently: Egyptian hieroglyphics, cuneiform, Chinese characters, etc. Scribes and readers then learned thousands of symbols, which necessarily was restricted to a small part of society. The great democratizing invention of alphabetic writing, which dramatically improved information handling (and irreversibly changed how we speak, hear, and remember), was done only once in history. All known alphabets derive from that seminal (Semitic) script. The idea was to make writing both to convey meaning and to reproduce (extremely poorly!) the way speech sounds. Of course, the old logograms have some phonetic component, generally based on the rebus principle, mostly representing the names of foreign rulers (Putin $=$ put $+$ in). But the alphabet makes a complete transition, using phonograms instead of logograms. The way we hear is related to the notion of phonemes. Linguists define the phoneme as the smallest acoustic unit that makes a difference in meaning. Their exact number in different languages is subject to disagreement, but generally there are a few dozen, which is comparable to the number of letters in the alphabet. Another interesting question is how we recognize words in a speech, which is essentially a running stream of sound—apparently, rhythm plays the

"The prophecy says they will turn back to hieroglyphs in the twenty-first century."

leading role. Coming full circle, the new age has brought two new hieroglyphic encodings: emoticons and emoji.

The statistics of words and their meanings is the subject of much research and still poorly understood. It was found empirically that if one ranks words by the frequency of their appearance in texts, then the frequency decreases with the rank by a power law: $p_r = Cr^{-\alpha}$ (Zipf 1949). In first place is *the* with 7%, followed by *of* with 3.5%, *and* with 1.7%, etc. Most data give $\alpha \simeq 1$. Such a decrease is remarkably slow—rare words appear quite frequently. Moreover, $\sum_{r=1}^{V} r^{-1}$ grows with the vocabulary size $V$. Another empirical finding is that the vocabulary itself grows approximately as a square root of the text length, $V \propto \sqrt{N}$ (Herdan 1960, Heap 1978). Word length distributions also keep changing with the sample size (shifts to longer words). Such nonsaturation is a nontrivial feature of the word statistics.

Probably the simplest model that gives the Zipf law is random typing: all letters plus the space are taken with equal probability (Miller 1957, Wentian Li 1992). Then any word with the length $L$ is flanked by two spaces and has the probability $P_i(L) = (M+1)^{-L-2}/Z$, where $i = 1, 2, \ldots, M^L$ and $M$ is the alphabet size. The normalization factor is $Z = \sum_L M^L(M+1)^{-L-2} = (M+1)^2/M$. Since the probability decreases with $L$, so does the rank, so that $r(L)$ is a growing function: $0 < r(1) \leq M$, $M < r(2) \leq M(M+1)$. Generally, the number of words with a length not exceeding $L$ is $\sum_{i=1}^{L} M^i$, so that $r(L)$ of any $L$ word satisfies the inequality

$$M(M^{L-1} - 1)/(M - 1) = \sum_{i=1}^{L-1} M^i < r(L) \leq \sum_{i=1}^{L} M^i = M(M^L - 1)/(M - 1),$$

which can be written as $P_i(L) < C[r(L) + B]^{-\alpha} \leq P_i(L - 1)$ with $\alpha = \log_M(M+1)$, $B = M/(M - 1)$, and $C = B^\alpha/M$. Note that $\alpha > 1$ so that the distribution is normalizable. For $M = 26$, we have $\alpha = 1.01$, and in the limit of large alphabet size, $M \gg 1$, we obtain

$$P(r) \propto (r + 1)^{-1}. \tag{A.11}$$

This asymptotic actually takes place for wide classes of letter distributions, not necessarily equiprobable. Does then the Zipf law trivially appear because both the number of words (inverse probability) and rank increase exponentially with the word size? The answer is negative because the number of distinct words of the same length in real language is not exponential in length and is not even monotonic (though on average the brevity law holds: the frequency of a given word decays exponentially with its length). Moreover, if we estimate the maximal word length for a finite (typical!) $N$ text by $P(L_{max})N = 1$, then the random typing model gives the highest rank, i.e., vocabulary $V = r(L_{max})$, growing linearly with the length, $N/M^4 < V < N/M^3$, which does not correspond to reality. Random models also give wrong lexical density (the number of words differing from a given word by one letter), which is very important for errors in transmission and comprehension.

It is reassuring that our texts are statistically distinguishable from those produced by an imaginary monkey with a typewriter. Besides, words mean something. The number of meanings (counted from the number of dictionary entries for a word) grows approximately as the square root of the word frequency: $m_i \propto \sqrt{P_i}$. Meanings correspond to objects of reference having their own probabilities, and it might be that the language combines these objects into groups whose sizes are proportional to the mean probability of the group $p_i$, so that $P_i = m_i p_i \propto m_i^2$.

A way of *interpreting* statistical distributions is to look for the variational principle they satisfy. What we mostly do in this book and what most statisticians do most of the time is to look for a conditional entropy maximum and minimize a two-term functional. In this case, one may require maximal information transferred with the least effort. It is tempting to suggest that the word distributions appear due to the balance between minimizing the efforts of writers and readers, or speakers and listeners. Writers and speakers would minimize their effort by having one word meaning everything and appearing with the probability of one. On the other end, the difficulty of perception is proportional to the depth of the memory keeping the context, needed, in particular, for choosing the right meaning. Readers and listeners then prefer a lot of single-meaning words. The rate of information transfer is $S = -\sum_r P(r) \log P(r)$. The effort must be higher for less common words, that is, it must grow with the rank. It is natural to assume that the effort is proportional to the word length $L$; then it is logarithmic in $r$. The mean effort is then $W = \sum_r P(r) \log r$. Looking for the minimum of $S - \lambda W$, we obtain $P(r) \propto r^{-\lambda}$ (Mandelbrot 1962). The Zipf law corresponds to $\lambda = 1$, when goals and efforts are balanced. So far, no convincing optimization scheme giving different features of word statistics has been found.

Some statistical laws of language probably could be better understood by treating it as a signal system. The information that changes the behavior of a receiver is referred to as a *signal*. For example, some conversations are

not exchanges of information but rather mating games, with multiple syn-
onyms and meanings as verbal plumage (Miller 2011). The information-
versus-manipulation dilemma is also at the heart of the fascinating study of
animal signals, where reliability and cost are central themes. For example, if
a signal indicates the intention to attack, the population of honest signalers
seems unstable to the invasion of dishonest signalers who drive opponents from
precious resources without any intention or ability to attack; eventually, the
majority will be dishonest and the signal meaningless (Maynard Smith and
Price 1973). One hypothetical resolution is the handicap principle, which suggests that some signals are too metabolically costly to fake: a fitness-lowering
peacock tail is a reliable signal of greater fitness (Zahavi 1975). Another mechanism involves mutual benefits between the signaler and receiver: a bright color
of a toxic animal benefits both it and a potential predator. Nontoxic animals can
use mimicry, but it is effective only if rare, as any unreliable signal. A quantitative
description of the evolution and stable equilibria of signal systems is a subject
of much interest. Even more tantalizing is the question of how any system of signals and codes gets started. After all, an absence of signaling also seems a stable
equilibrium: no signals means no ability to respond, which means no need to
signal. Related are the studies of the appearance and evolution of cooperation.
Such problems are generally treated in evolutionary game theory, which considers optimization strategies (like we do in sections 3.6 and A.6) but in more
general situations where the result depends on the behavior of other players.

## A.4    Ising Model of the Brain

As was mentioned at the end of section 3.1, the activity of a network of neurons could be described by one-bit variables, $\sigma_i = \pm 1$ (active or inactive). The
probability distribution that maximizes entropy under the given set of mean
activities $\langle \sigma_i \rangle$ and pairwise correlations $\langle \sigma_i \sigma_j \rangle$ is that of an Ising model (3.3):

$$\rho(\{\sigma\}) = Z^{-1} \exp \left[ \sum_i h_i \sigma_i + \frac{1}{2} \sum_{i<j} J_{ij} \sigma_i \sigma_j \right]. \tag{A.12}$$

One can also measure some multicell correlations and check how well they
agree with those computed from (A.12). Despite apparent patterns of collective behavior involving many neurons, it turns out to be enough to account
for pairwise correlations to describe the statistical distribution remarkably
well (Schneidman, Berry, Segev, and Bialek 2006). This is also manifested
by the entropy changes: measuring double, triple, and multicell correlations
imposes more restrictions and lowers the entropy maximum. One then checks

that accounting for pairwise correlations changes the entropy significantly, while accounting for further correlation changes the entropy relatively little. The practical sufficiency of pairwise correlations provides an enormous simplification, which may be important not only for our description but also for the brain itself. The reason is that our mind actually develops and constantly modifies its own predictive model of probability needed in particular to accurately evaluate new events for their degree of surprise, as described in section 3.4. The dominance of pairwise interactions means that learning rules based on pairwise correlations could be sufficient to generate nearly optimal internal models to accurately evaluate probabilities. Side remark: We should not think that what is encoded from sensors into electrical neuron activity is then "decoded" inside the brain. Whatever it is, the brain is not a computer.

It is interesting how the entropy scales with the number of interacting neurons $N$. The entropy of noninteracting (or nearest-neighbor interacting) neurons is extensive, that is, proportional to $N$. The data show that $J_{ij}$ are nonzero for distant neurons as well. That means that the entropy of an interacting set is lower at least by the sum of the mutual information terms between all pairs of cells. The negative contribution is thus proportional to the number of interacting pairs, $N(N-1)/2$, that grow faster with $N$, at least when it is not too large. One can estimate from low $N$ data a "critical" $N$ when the quadratic term is expected to turn entropy into zero. That critical $N$ corresponds well to the empirically observed sizes of the clusters of strongly correlated cells. The lesson: Even when pairwise correlations are weak, sufficiently large clusters can be strongly correlated. It is also important that the interactions $J_{ij}$ have different signs so that frustration can prevent the freezing of the system into a single state (like ferromagnetic or antiferromagnetic). Instead, there are multiple states that are local minima of the effective energy function, as in spin glasses.

## A.5   Applying the Infomax Principle

The maximal-capacity approach described in section 3.3 turns out to be quite useful for image and speech recognition by iterative algorithms. One chooses some form of the response function $y = g(x, w)$ characterized by the parameter $w$ and finds the optimal value of $w$ using an "online" stochastic gradient ascent learning rule, giving the change of the parameter:

$$\Delta w \propto \frac{\partial}{\partial w} \ln \left( \frac{\partial g(x, w)}{\partial x} \right). \tag{A.13}$$

Of course, the eye or camera provides not a single input signal but the whole picture. Let us consider $N$ inputs and outputs (neurons/channels). An input vector $\mathbf{x} = (x_1, \ldots, x_N)$ is transformed into the output vector $\mathbf{y}(\mathbf{x})$ one to one, that is, $\det[\partial y_i / \partial x_k] \neq 0$. The multivariate probability density function of $y$ is as follows:

$$\rho(\mathbf{y}) = \frac{\rho(\mathbf{x})}{\det[\partial y_i / \partial x_k]}. \tag{A.14}$$

Making it flat (distributing outputs uniformly) for maximal entropy is not straightforward now. In one dimension, it is enough to follow the gradient to arrive at the closest extremum, but there are many possible paths to different mountain summits. Maximizing the total mutual information between input and output, which requires maximizing the output entropy, is often (but not always) achieved by first minimizing the mutual information between the output components. For two outputs, we may start by maximizing $S(y_1, y_2) = S(y_1) + S(y_2) - I(y_1, y_2)$, that is, minimizing $I(y_1, y_2)$. If we are lucky and find encoding in terms of independent components, then we choose for each component the transformation (3.8), which maximizes its entropy, making the probability flat; see exercise 2.10.

*Maximizing* the mutual information between input and output by first *minimizing* the mutual information between the components of the output is particularly useful for natural signals, where most redundancy comes from strong correlations (like that of the neighboring pixels in visuals). Also, finding an encoding in terms of the least dependent components is important by itself for its cognitive advantages. For example, such encoding generally facilitates pattern recognition. In addition, presenting and storing information in the form of independent (or minimally dependent) components is important for associative learning done by brains and computers. To learn a new association between two events $A$ and $B$, the prior joint probability $P(A, B)$ is needed. For correlated $N$-dimensional $A$ and $B$, one needs to store $N \times N$ numbers, but only $2N$ numbers for quantities uncorrelated (until the association occurs).

Another cognitive task of identifying independent components is the famous "cocktail party problem" posed by security services: $N$ microphones (flies on the wall) record $N$ people speaking simultaneously, and we need the program to separate them—the *blind separation* problem. A less sinister aspect of such an auditory scene analysis is when you try to follow one of the conversations in a noisy hall—it is easy when you look at the speaking person and it is difficult when you don't. Formally, the problem is as follows: Uncorrelated sources $s_1, \ldots, s_N$ are mixed linearly by an unknown matrix $\hat{A}$.

All we receive are the $N$ superpositions of them, $x_1, \ldots, x_N$. The task is to recover the original sources by finding a square matrix $\hat{W}$, which is the inverse of the unknown $\hat{A}$, up to permutations and rescaling. Closely related is the *blind de-convolution* problem (see, e.g., Bell and Sejnowski 1995): a single unknown signal $s(t)$ is convoluted with an unknown filter, giving a corrupted signal $x(t) = \int a(t - t')s(t')\, dt'$, where $a(t)$ is the impulse response of the filter. This time the signal is corrupted not by other signals but by the time-shifted version of itself. The task is to recover $s(t)$ by integrating $x(t)$ with the inverse filter $w(t)$, which we need to find by learning procedure. Upon discretization, $s, x$ are turned into $N$ vectors and $w$ into an $N \times N$ matrix, which is lower triangular because of causality: $w_{ij} = 0$ for $j > i$ and the diagonal values are all the same, $w_{ii} = \bar{w}$. The determinant in (A.14) is simplified in this case. For $\mathbf{y} = g(\hat{w}\mathbf{x})$, we have $\det[\partial y(t_i)/\partial x(t_j)] = \det \hat{w} \prod_i^N y'(t_i) = \bar{w}^N \prod_i^N y'(t_i)$. One then applies some variant of the gradient ascent method to minimize mutual information.

Ideally, we wish to find the (generally stochastic) encoding $\mathbf{y}(\mathbf{x})$ that achieves the absolute minimum of the mutual information $\sum_i S(y_i) - S(\mathbf{y})$. One way to do that is to minimize the first term while keeping the second one, that is, under the condition of the fixed entropy $S(\mathbf{y}) = S(\mathbf{x})$. In general, one may not be able to find such encoding without any entropy change, $S(\mathbf{y}) - S(\mathbf{x})$. In such cases, one defines a functional that grades different codings according to how well they minimize *both* the sum of the entropies of the output components *and* the entropy change. The simplest energy functional for statistical independence is then

$$E = \sum_i S(y_i) - \beta[S(\mathbf{y}) - S(\mathbf{x})] = \sum_i S(y_i) - \beta \ln \det[\partial y_i/\partial x_k]. \quad \text{(A.15)}$$

A coding is considered to yield an improved representation if it possesses a smaller value of $E$. The choice of the parameter $\beta$ reflects our priorities— whether statistical independence or an increase in indeterminacy is more important.

Maximizing information transfer and reducing the redundancy between the units in the output is applied in practically all disciplines that analyze and process data, from physics and engineering to biology, psychology, and economics. Within the general infomax domain, this specific technique is called independent component analysis. More sophisticated schemes employ not only mutual information but also interaction information (2.32). The redundancy reduction is usually applied after some procedure of eliminating noise.

This is because our gain function provides equal responses for probable and improbable events, but the latter can be mostly due to noise, which thus needs to be suppressed. Moreover, if input noises are uncorrelated, they can get correlated after coding. More generally, it is better to keep some redundancy for corrections and checks when dealing with noisy data.

## A.6　Exploration or Exploitation: Index Strategy

Devising gambling strategies in section 3.6, we took the probabilities $\{p_i\}$ as given. But more often one needs to play the game to learn the chances. As one plays, incurs some gains and losses, and collects some information, one needs to strike the right balance between the exploitation of existing information to maximize the gain and an exploration for new information. There is a broad class of the so-called sequential allocation problems encompassing the design of clinical trials, adaptive routing, job scheduling, gambling, and military logistics. Optimal for all of them is the remarkable index strategy, which we first illustrate using the simple problem of scheduling jobs: Job $i$ takes time $t_i$ and, on completion, gives reward $r_i$. It is important that later rewards are $\gamma^t$ less valuable, where the discount factor $0 < \gamma < 1$. To maximize the total discounted reward, we do $i$ before $j$, if $r_i \gamma^{t_i} + r_j \gamma^{t_i+t_j} > r_j \gamma^{t_j} + r_i \gamma^{t_i+t_j}$. Taking $r_i$ terms to the left and $r_j$ terms to the right, we can present this as an inequality for the job indices:

$$v_i = \frac{\gamma^{t_i}}{1 - \gamma^{t_i}} r_i > v_j = \frac{\gamma^{t_j}}{1 - \gamma^{t_j}} r_j.$$

So we can compute the index $v_i$ for each job independently and schedule the jobs in decreasing order of the indices.

It is remarkable that there exists an index strategy for problems where each option has a random element with statistics initially unknown to us. Let us play a so-called multiarmed bandit game, where we can only make one bet at a time, choosing among several options (arms of slot machines). Each arm has some probability of winning, $0 \leq s_i \leq 1$, and gives the same reward: $r = 1$ if you win and 0 if you lose. At the start, we do not know the probabilities of winning, $s_i$, so we take a uniform prior: $P(s_i) = 1$. We play each arm several times and compute the posterior distribution by Bayes' formula. If we encounter $w_i$ wins and $l_i$ losses, then for every value of $s_i$, the posterior probability is the binomial distribution of $w_i, l_i$ happening:

$$P(s_i) = s_i^{w_i} (1 - s_i)^{l_i} \frac{(w_i + l_i + 1)!}{w_i! l_i!}. \tag{A.16}$$

Upon further trials with $l'$ losses and $w'$ wins, the distribution is multiplied by $s_i^{w'}(1-s_i)^{l'}$; that is, it preserves its form, renormalizing parameters. A Bayesian update for every arm is equivalent to a random walk in a positive direction on a two-dimensional lattice $(w_i, l_i)$. Each of these lattice points is a state of a Markov process with the one-step vector of transition probabilities $P = \{w_i/(w_i + l_i), l_i/(w_i + l_i)\}$.

We need a strategy that maximizes the sum of the discounted rewards: the expected value of the sum $r_0 + \gamma r_1 + \gamma^2 r_2 + \dots$ . Even though the total number of steps is potentially infinite, the discount factor introduces an effective horizon $\simeq (1 - \gamma)^{-1}$. The powerful statement that we give without proof is that the optimal strategy is to play at each step the arm with the maximal index $v_i$ (Gittins 1979). The index is the ratio of the expected sum of rewards to the discounted time, under the assumption that playing the arm will be terminated in the future after $\tau$ steps:

$$v(l_i, n_i, t) = \sup_{\tau > 0} \frac{\sum_{k=0}^{\tau-1} \gamma^k \langle r_{t+k-1} \rangle}{\sum_{k=0}^{\tau-1} \gamma^k}. \tag{A.17}$$

Here

$$\langle r_{t+k-1} \rangle = \frac{w_i(t_i + k - 1)}{w_i(t_i + k - 1) + l_i(t_i + k - 1)}$$

is the expected reward at step $k$, and we sum the future rewards that one would obtain by choosing to play only the $i$th arm up to the stopping time $t + \tau$. The brackets denote the averaging over all the lattice paths with expectations based on the distributions (A.16) at every lattice point $w_i(t_i + k - 1), l_i(t_i + k - 1)$. We take the maximum over the number of future steps, which is variable since we admit the possibility of a switch to another arm. That supremum can be shown to be achieved; that is, the stopping time $\tau$ is finite because the discounted time in the denominator of (A.17) grows with $\tau$. Denote $L = v/(1 - \gamma)$, then

$$L = \frac{1}{1 - \gamma^\tau} \sum_{k=0}^{\tau-1} \gamma^k \langle r_{t+k-1} \rangle = \sum_{k=0}^{\tau-1} \gamma^k \langle r_{t+k-1} \rangle + \gamma^\tau L.$$

That formula means that the lump sum $L$ either now or after some optimal number of further rewards is an equally good alternative. One then obtains $L$ (numerically) as a maximal reward, which is a fixed point that does not change upon one step.

The game thus proceeds as follows: At the beginning, all indices are equal. We start from an arbitrary arm and play it until the number of losses makes its index less than other ones, then we switch to another one, etc. After a while, all arms are played many times with switches occurring when enough losses are encountered. In the limit $l_i + w_i \to \infty$, the probability shrinks to $P(s_i) = \delta(s_i - p_i)$, where $p_i = \lim_{l_i+w_i \to \infty} w_i/(w_i + l_i)$, the mean reward is $r_0 = p_i$, and the evident optimal strategy is to choose the arm with the highest $p_i$, that is, $v_i = r_0 = p_i$. Generally, the finite-time index is larger than its infinite-time asymptotic, accounting for the possibility that the actual probability is larger than the observed one. As we play an arm, its distribution (A.16) is getting more and more narrow and the index decreases, which makes it possible to switch to another arm. Switching arms provides the possibility of exploration and obtaining new information.

## A.7   Memory Effects in Particle Collisions

Let us understand how corrections to the Boltzmann kinetic equation can diverge with system size. To estimate the correction proportional to the squared concentration, one needs to estimate the probability of a repeated collision. For simplicity, assume that a given particle is scattered by randomly placed, distant fixed scatterers with concentration $\rho$. The simplest case of a repeated collision is shown in the figure: the particle is reflected by scatterer 2 to collide with 1 the second time.

The probability of such an event is proportional to $\rho^2$ times the solid angle of scatterer 2 as seen from 1 times the solid angle of scatterer 1 as seen from 2. To get the whole probability, one needs to integrate over all possible positions of scatterer 2. Let us show that such a probability diverges at large distances in 2D and converges in 3D. The solid angles of the scatterers are proportional to $R^{1-d}$, where $R$ is the distance between 1 and 2, considered to be much larger than the size of the scatterer. The probability of finding the second scatterer at such a distance is proportional to $R^{d-1}dR$. The total probability of a triple collision giving a $\rho^2$ correction is thus given by the integral $\int R^{d-1} R^{2(1-d)} dR = \int R^{1-d} dR$, which diverges at $d \leq 2$. Moreover, the correction proportional to $\rho^n$ involves integration over $n - 1$ positions of distant scatterers and $n$ solid angles, $R^{(n-1)d+n(1-d)}$, which diverges for all $n \geq d$. That shows how the memory of past collisions and correlations accumulates. In particular, a $\rho^3$ correction is determined by the probability of returning to 1 after two consequent scatterings by 2 and 3 (not shown); such a probability integrated over all possible positions of 2 and 3 could diverge in 3D.

Remarkably, all the divergences cancel in thermal equilibrium due to detailed balance, so that the equation of state has a regular virial expansion. For example, the pressure is an analytic function of density for dilute gases. For nonequilibrium states, the cancellations are generally absent and one has to deal with the divergences. In particular, that takes place if one tries to apply the expansion over density to kinetic coefficients, like diffusivity, conductivity, or viscosity. Of course, the divergences appear because the "naive" virial expansion allows particles to travel arbitrarily long distances between collisions. One must account for the collective effects that impose the large-distance cut-off as the mean free path proportional to $1/\rho$. That requires resummation and brings logarithmic dependency of $\rho$. As a result, kinetic coefficients and other nonequilibrium properties are nonanalytic functions of density.

## A.8   Baker's Map

Here we present a toy model that is able to describe both the mixing of area-preserving flows and the fractalization of compressible flows.

*Area-preserving map and mixing*   Consider first the area-preserving transformation: an expansion in the $x$ direction and a contraction in the $y$ direction, arranged so that the unit square is mapped onto itself at each step. The transformation consists of two steps: First, the unit square is contracted in the $y$ direction and stretched in the $x$ direction by a factor of two. The unit square becomes a rectangle, $0 < x < 2$, $0 < y < 1/2$. Next, the rectangle is cut vertically in the middle and the right half is put on top of the left half to recover a square.

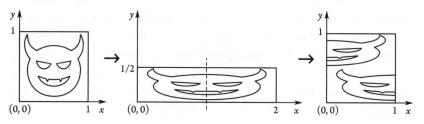

This is how bakers prepare long, thin strips of pasta. If we consider two initially close points, after $n$ such steps the distance along $x$ and $y$ will be multiplied respectively by $2^n = e^{n \ln 2}$ and $2^{-n} = e^{-n \ln 2}$. It is then easy to see, without a lot of formalities, that there are two Lyapunov exponents corresponding to the discrete time $n$. The one connected to the expanding direction is $\lambda_+ = \ln 2$. The other connected to the contracting direction is $\lambda_- = -\ln 2$. For the forward time operation of the baker's transformation, the expanding direction is along the $x$ axis and the contracting direction is along the $y$ axis. If one considers the time-reversed motion, the expanding and contracting directions change places. Therefore, for the forward motion, nearby points separated only in the $y$ direction approach each other exponentially rapidly with the rate $\lambda_- = -\ln 2$. In the $x$ direction, points separate exponentially with $\lambda_+ = \ln 2$. The sum of the Lyapunov exponents is zero, which reflects the fact that the baker's transformation is area-preserving.

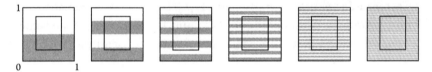

FIGURE A.1. Iterations of a baker's map.

Let us argue now that the baker transformation is mixing, that is, spreading the measure uniformly over the whole phase space. If a measure is initially concentrated in any domain, as in the gray areas in figure A.1, after a sufficiently long time the domain is transformed into a large number of very thin, horizontal strips of length unity, distributed more and more uniformly in the vertical direction. Eventually, any set in the unit square will have the same fraction of its area occupied by these little strips of pasta as any other set. This is the indicator of a mixing system. If we add to that a small coarse-graining, at a sufficiently long time it blurs our measure to a constant one. We conclude that a sufficiently smooth initial distribution function defined on the unit square will approach a uniform (microcanonical) distribution. The baker's map is area-preserving and does not change entropy. When we add repeated coarse-graining along with the evolution, then the entropy grows and eventually reaches the maximum, which is the logarithm of the phase volume, as must be the case for the microcanonical distribution.

One can encode any point on the square using the binary code of the expansion $x = \sum_{k=0}^{\infty} a_k 2^{-k-1}$, $y = \sum_{i=1}^{\infty} b_{-i} 2^{-i}$, where all $a_k, b_{-i}$ are either 1 or 0. Simply, $a_1$ encodes in which half, $a_2$ encodes in which quarter within the half, etc. Encoding any point as a bi-infinite string, $(x, y) = \ldots b_{-3} b_{-2} b_{-1} . a_0 a_1 a_2 \ldots$, one finds out that the baker's map shifts the point . one step to the right. Using such so-called symbolic dynamics, one can analyze the mixing properties of maps.

To avoid the impression that cutting and gluing of the baker's map are necessary for mixing, consider a smooth model with a similar behavior. Namely, consider a unit two-dimensional torus, that is, a unit square with periodic boundary conditions, so that all distances are measured modulo 1 in the $x$ and $y$ directions. We map $x' = ax + by$ and $y' = cx + dy$. The action of such a toral map is shown in figure A.2; it maps a unit torus into itself if $a, b, c, d$ are all integers. The eigenvalues $\lambda_{1,2} = (a+d)/2 \pm \sqrt{(a-d)^2/4 + bc}$ are real when $(a-d)^2/4 + bc \geq 0$. For the transform to be area-preserving, the product of the eigenvalues must be unity: $\lambda_1 \lambda_2 = ad - bc = 1$. In a general case, one eigenvalue is larger than unity and one is smaller, which corresponds respectively to positive and negative Lyapunov exponents $\ln \lambda_1$ and $\ln \lambda_2$.

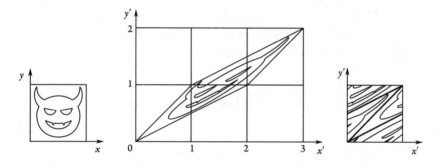

FIGURE A.2. Toral map.

*Compressible map and fractalization*    To illustrate the entropy decay and fractalization in compressible flows, we consider a slight generalization of the baker's map, expanding one region and contracting another, keeping the whole area:

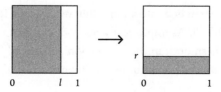

The transformation has the form

$$x' = \begin{cases} x/l & \text{for } 0 < x < l \\ (x-l)/r & \text{for } l < x < 1 \end{cases},$$

$$y' = \begin{cases} ry & \text{for } 0 < x < l \\ r+ly & \text{for } l < x < 1 \end{cases}, \tag{A.18}$$

where $r + l = 1$. The Jacobian of the transformation is not identically equal to unity when $r \neq l$:

$$J = \left| \frac{\partial(x',y')}{\partial(x,y)} \right| = \begin{cases} r/l & \text{for } 0 < x < l \\ l/r & \text{for } l < x < 1 \end{cases}. \tag{A.19}$$

If $l > 1/2$, then $r = 1 - l < l$, so that $J < 1$ in the shaded region where $x < L$ and $J > 1$ in the white region where $x > L$. Of course, the mean Jacobian $\bar{J} = r + l$ is unity, since we always occupy the same unit square. Like in the treatment of the incompressible baker's map, consider two initially close points. If during $n$ steps the points find themselves $n_1$ times in the region $0 < x < l$ and $n_2 = n - n_1$ times inside $l < x < 1$, then the distances along $x$ and $y$ will be multiplied respectively by $l^{-n_1} r^{-n_2}$ and $r^{n_1} l^{n_2}$. Taking the log and the limit, we obtain the Lyapunov exponents:

$$\lambda_+ = \lim_{n\to\infty} \frac{1}{n} \ln \frac{\delta x(n)}{\delta x(0)} = \lim_{n\to\infty} \left[ \frac{n_1}{n} \ln \frac{1}{l} + \frac{n_2}{n} \ln \frac{1}{r} \right] = -l\ln l - r\ln r,$$

$$\tag{A.20}$$

$$\lambda_- = \lim_{n\to\infty} \frac{1}{n} \ln \frac{\delta y(n)}{\delta y(0)} = \lim_{n\to\infty} \left[ \frac{n_1}{n} \ln r + \frac{n_2}{n} \ln l \right] = r\ln l + l\ln r. \tag{A.21}$$

The sum of the Lyapunov exponents, $\lambda_+ + \lambda_- = (l-r)\ln(r/l) = \overline{\ln J}$, is nonpositive and is zero only for $l = r = 1/2$. Again, the convexity of the logarithmic function means that $\overline{\ln J} \leq \ln \bar{J} = 0$. The volume contraction means that the expansion in the $\lambda_+$ direction proceeds slower than the contraction in the $\lambda_-$ direction. After $n$ iterations of the map, a square having initial side $\delta \ll L$ will be stretched into a long, thin rectangle of length $\delta \exp(n\lambda_+)$ and width $\delta \exp(n\lambda_-)$. Asymptotically our strips of pasta concentrate on a fractal set, which is smooth in the $x$ direction and fractal in the $y$ direction. That gives two terms in the noninteger dimensionality, $d_f = 1 + \lambda_+/|\lambda_-|$; see (5.21).

Let us now use our model to derive the relation between the probabilities of entropy increase and decrease:

$$P^\dagger(-\Delta S) = P(\Delta S)e^{-\Delta S}. \tag{A.22}$$

Here $P^\dagger$ refers to a time-reversed process. At every step, the volume contraction factor is the Jacobian of the transformation: $J = r/l$ for $x \in (0, l)$ and $J = l/r$ for $x \in (l, 1)$. A longtime average rate of the entropy production, $\overline{\ln J} = (l - r) \ln(r/l)$, is the volume contraction rate of a fluid element. However, during a finite time $n$, there is always a finite probability to observe an expansion of an element. This probability must decay exponentially with $n$, and there is a universal law relating relative probabilities of the extraction and contraction. If during $n$ steps a small rectangular element finds itself $n_1$ times in the region $0 < x < l$ and $n_2 = n - n_1$ times inside $l < x < 1$, then its sides along $x$ and $y$ will be multiplied respectively by $l^{-n_1} r^{-n_2}$ and $r^{n_1} l^{n_2}$. The volume contraction factor for such an $n$ sequence is $(l/r)^{n_2 - n_1}$, and its log is $\Delta S = n \ln J = n_1 \ln \frac{r}{l} + n_2 \ln \frac{l}{r}$. The probability of the sequence is $P(\ln J) = l^{n_1} r^{n_2}$. The opposite sign of $\ln J$ takes place, for instance, in a time-reversed sequence. Time reversal corresponds to the replacement $x \to 1 - y, y \to 1 - x$, that is, the probability of such a sequence is $P(-\ln J) = r^{n_1} l^{n_2}$. Therefore, denoting the entropy production rate $\sigma = -\ln J$, we obtain the universal probability independent of $r, l$:

$$\frac{P(\Delta S)}{P(-\Delta S)} = \frac{P(\sigma)}{P(-\sigma)} = \left(\frac{l}{r}\right)^{n_2 - n_1} = e^{n\sigma} = e^{\Delta S}. \tag{A.23}$$

## A.9    Multidimensional Renormalization Group

Here we qualitatively describe the renormalization group flow for the Ising model in higher dimensions. Recall that we consider the Gibbs distribution $\rho\{\sigma_i\} = Z^{-1} \exp\left(-K \sum_{i,j} \sigma_i \sigma_j\right)$, where $\sigma_i = \pm 1$, $i, j$ are the nearest neighbors, and $K = 1/T$ is the parameter to be renormalized upon passing from spins to blocks.

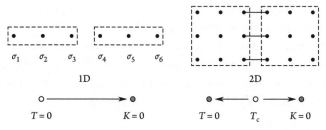

*Critical point and phase transition*    We have seen in section 5.5 that renormalization in 1D is according to the law $K' = \tanh^{-1}(\tanh^3 K)$. Since $\tanh K < 1$, then $K' < K$, i.e., $K$ decreases under RG transformation. That means that the zero-temperature fixed point $(K = \infty)$ is unstable in 1D. Note, however, that in the low-temperature region, $K$ decreases slowly so that it does not change in the main order: $K' = K - \text{const} \approx K$. This can be readily interpreted: the interaction between $m$ blocks is mediated by their boundary spins, which all point in the same direction, $K' \approx K\langle\sigma_3\rangle_{\sigma_2=1}\langle\sigma_4\rangle_{\sigma_5=1} \approx K$ (by the same token, at high temperatures $\langle\sigma\rangle \propto K$ so that $K' \propto K^3$). However, in $d$ dimensions, there are $m^{d-1}$ spins at the block side so that $K' \approx m^{d-1}K$ as $K \to \infty$. For $m = 3$ and $d = 2$, we have $K' \approx 3K$; see the right panel of the preceding figure. That means that $K' > K$, that is, the low-temperature fixed point is stable at $d > 1$. On the other hand, the paramagnetic fixed point $K = 0$ is stable too, so there must be an unstable fixed point in between at some $K_c$ that corresponds to a critical temperature $T_c$.

In contrast to the case of summing random numbers in section 5.5, we are interested now in an unstable fixed point, because it separates regions between two qualitatively different large-scale behaviors—ordered and disordered. At a finite temperature, there are always ordered and disordered domains of different scales. At $T > T_c$, looking at larger and larger domains we find them less and less correlated with each other. At $T < T_c$, the mean spins of larger and larger domains are more and more correlated with each other.

Another new aspect of multidimensional spin systems is the need to consider RG flows not only along the line of $K$ values but also in multidimensional parameter spaces. As seen in the left panel of figure A.3, summing over a corner spin $\sigma$ produces diagonal coupling between blocks. In addition to $K_1$, which describes an interaction between neighbors, we need to introduce another parameter, $K_2$, to account for a next-nearest-neighbor interaction. In fact, RG generates all possible further couplings so that it is a flow in an infinite-dimensional $\mathbf{K}$ space. An unstable fixed point in this space determines critical behavior. The dimensionality of the attractor is determined by the number of unstable directions. To be at criticality, displacements in the unstable directions must be kept at zero, which requires tuning the respective parameters. We know, however, that we need to control a finite number of parameters to reach a phase transition; for Ising at zero external field and many other systems, it is a single parameter, temperature. For all such systems, RG flow has only one unstable direction; all the rest must be contracting stable directions, like the projection on the $K_1, K_2$ plane shown in the right panel of figure A.3. The line

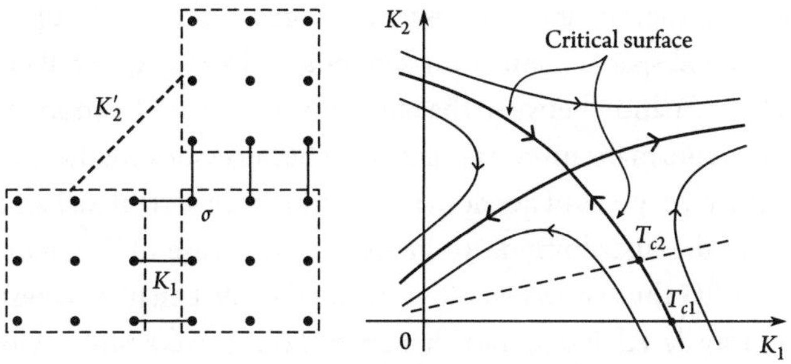

FIGURE A.3. Left: Next-nearest-neighbor coupling $K_2$ due to a corner spin. Right: Renormalization group flow with an unstable fixed point.

of points attracted to the fixed point is the projection of the critical surface, so called because the long-distance properties of each system corresponding to a point on this surface are controlled by the fixed point. The critical surface is a separatrix, dividing points that flow to high $T$ (paramagnetic) behavior from those that flow to low $T$ (ferromagnetic) behavior at large scales.[1]

We can now understand why physicists are so interested in the critical surface, where the fixed point is actually stable and attractive. That picture of the RG flow explains the universality of long-distance critical behavior: different physical systems (in different regions of the parameter **K** space) flow to the same fixed point, that is, have the same statistics of large-scale correlations and fluctuations. For example, changing the temperature in a system with only nearest-neighbor coupling, we move along the line $K_2 = 0$. The point where this line meets the critical surface defines $K_{1c}$ and respectively $T_{c1}$. At that temperature, the large-scale behavior of the system is determined by the RG flow, i.e., by the fixed point. In another system with nonzero $K_2$ and changing $T$, we move along some other path in the parameter space, indicated by the dashed line in the right panel of figure A.3. The intersection of this line with the critical surface defines some other critical temperature $T_{c2}$. However, the long-distance properties of this system at that temperature are determined by the same fixed point.

*Irreversibility of RG flow* It seems reasonable to expect the irreversibility of the renormalization group since it is a way of forgetting. Yet it is far from

1. We mention in passing that, in dimensions $d > 4$, the block-spin renormalization of the Ising-class models leads to asymptotic Gaussian distribution of the coarse-grained magnetization: $\ln \rho(\eta) \propto -|\nabla \eta|^2$.

trivial to find entropic characteristics that change monotonically upon RG in a multidimensional space. Eliminating modes step-by-step generally decreases the mutual information between the remaining degrees of freedom. However, rescaling and renormalization may increase it because some of the information about eliminated degrees of freedom is stored in the renormalized values of the parameters of the distribution. An increase or decrease of the mutual information upon RG thus shows whether the large-scale behavior is respectively ordered or disordered. It does not characterize the irreversibility of forgetting.

In Section 5.5, we characterized information exchange in one dimension, looking at a single bond that separates two parts of a spin chain. Breaking a single bond in more than one dimension does not cause separation. In a 2D plane, one can consider a (finite) line $L$ and break the direct interactions between the degrees of freedom on the different sides of it. That is, we make a cut and ascribe to every point two (generally different) values on the opposite sides. The statistics of such a set are now characterized not by a vector of probability on the line but by a matrix $\rho_L$, similar to the density matrix in quantum statistics described in section 6.3. For that matrix, one defines von Neumann entropy as $S_L = -\text{Tr}\rho_L \log \rho_L$.

For long lines in short-correlated systems, that quantity can be shown to depend only on the distance $r$ between the endpoints and not on the form of a line connecting them (that is, information flows like an incompressible fluid). Moreover, this dependence is logarithmic at criticality (when we have fluctuations of all scales). To cancel nonuniversal terms depending on the microscopic detail, one defines the function $c(r) = rdS_L(r)/dr$, which is shown to be a monotonic zero-degree function, using the positivity of the mutual information (subadditivity of the entropy) between lines with $r$ and $r + dr$ (Zamolodchikov 1986, Casini and Huerta 2007). The same function changes monotonically under RG flow and in a fixed point takes a finite value equal to the so-called zero charge of the conformal field theory. The zero charge is a measure of relevant degrees of freedom that respond to boundary perturbations. It is even more difficult to introduce a proper intensive measure of information flow in dimensions higher than two; so far it is done in a quite model-specific way (see, e.g., Komargodski and Schwimmer 2011, Klebanov et al. 2011).

In looking for fundamental characteristics of order in fluctuating systems in higher dimensions, one can go even deeper. For instance, one can consider for quantum systems in $2 + 1$ dimensions the relative entanglement of three finite planar regions, $A, B, C$, all having common boundaries. As a quantum analog of the interaction information (2.32), one can introduce so-called topological entanglement entropy, $S_A + S_B + S_C + S_{ABC} - S_{AB} - S_{BC} - S_{AC}$. For some classes of systems, one can show that the terms depending on the boundary lengths cancel out; what remains (if anything) can be thus independent of the deformations of the boundaries, that is, characterizing the topological order, if it exists in the system (Kitaev and Preskill 2006).

## A.10   Brownian Motion

We consider the motion of a small particle in a fluid. The momentum of the particle, $\mathbf{p} = M\mathbf{v}$, changes because of collisions with the molecules. Thermal equipartition guarantees that the mean kinetic energy of the particle is the same as the energy of any molecule and equal to $T/2$. When the particle mass $M$ is much larger than the molecular mass $m$, the RMS particle velocity, $v = \sqrt{T/M}$, is small compared to the typical velocities of the molecules, $v_T = \sqrt{T/m}$. That allows one to write the force $\mathbf{f}(\mathbf{p})$ acting on the particle as a Taylor expansion in $\mathbf{p}$, keeping the first two terms, independent of $\mathbf{p}$ and linear in $\mathbf{p}$: $f_i(\mathbf{p}, t) = f_i(0, t) + p_j(t)\partial f_i(0, t)/\partial p_j(t)$ (we neglect the dependence of the force on the momentum at earlier times). Such expansion makes sense if the neglected third term is much less than the second one, but then the second term must be much smaller than the first one—what is the reason to keep both? The answer is that molecules hitting a standing particle produce a force whose average is zero. The mean momentum of the particle is zero as well. However, random force by itself would make the squared momentum grow with time, like the squared displacement of a random walker in section 4.2. To describe the particle in equilibrium with the medium, the force must be balanced by resistance, which is also provided by the medium: the particle meets more molecules in the direction it moves and loses its momentum to them. That resistance has a nonzero mean and must be described by the second term, which then may be approximated as $p_j \partial f_i/\partial p_j = -\gamma p_j \delta_{ij} = -\gamma p_i$. The dimensionality of $\gamma$ is 1/sec; it is the rate of the momentum loss due to the resistance of the liquid. We then obtain

$$\dot{\mathbf{p}} = \mathbf{f} - \gamma \mathbf{p}. \tag{A.24}$$

The solution of the linear equation (A.24) is

$$\mathbf{p}(t) = \int_{-\infty}^{t} \mathbf{f}(t')e^{\gamma(t'-t)}\,dt'. \tag{A.25}$$

We must treat the force $\mathbf{f}(t)$ as a random function since we do not track molecules hitting the particle. We assume that $\langle \mathbf{f} \rangle = 0$ and that $\langle \mathbf{f}(t') \cdot \mathbf{f}(t' + t) \rangle = 3C(t)$ decays with $t$ during the correlation time $\tau$, which is much smaller than $\gamma^{-1}$ because it takes many collisions to change the momentum of the particle. Since the integration time in (A.25) is of order $\gamma^{-1}$, then the condition $\gamma\tau \ll 1$ means that the momentum of a Brownian particle can be considered as a sum of many independent random numbers (integrals over intervals of order $\tau$), so it must have Gaussian statistics $\rho(\mathbf{p}) = (2\pi\sigma^2)^{-3/2} \exp(-p^2/2\sigma^2)$, where

$$\sigma^2 = \langle p_x^2 \rangle = \langle p_y^2 \rangle = \langle p_z^2 \rangle = \int_0^\infty C(t_1 - t_2)e^{-\gamma(t_1+t_2)} dt_1 dt_2$$

$$\approx \int_0^\infty e^{-2\gamma t} dt \int_{-2t}^{2t} C(t') dt' \approx \frac{1}{2\gamma} \int_{-\infty}^\infty C(t') dt'. \qquad (A.26)$$

On the other hand, equipartition guarantees that $\langle p_x^2 \rangle = MT$ so that we can express the friction coefficient via the correlation function of the force fluctuations (a particular case of the detailed balance):

$$\gamma = \frac{1}{2TM} \int_{-\infty}^\infty C(t') dt'. \qquad (A.27)$$

Displacement,

$$\Delta \mathbf{r}(t') = \mathbf{r}(t + t') - \mathbf{r}(t) = \int_0^{t'} \mathbf{v}(t'') dt'',$$

is also Gaussian with a zero mean. To get its second moment, we need the different-time correlation function of the velocities,

$$\langle \mathbf{v}(t) \cdot \mathbf{v}(0) \rangle = (3T/M) \exp(-\gamma|t|), \qquad (A.28)$$

which can be obtained from (A.25). Note that friction makes velocity correlated on a longer timescale than force. That gives

$$\langle |\Delta\mathbf{r}|^2(t') \rangle = \int_0^{t'} dt_1 \int_0^{t'} dt_2 \langle \mathbf{v}(t_1)\mathbf{v}(t_2) \rangle = \frac{6T}{M\gamma^2}(\gamma t' + e^{-\gamma t'} - 1).$$

The mean squared distance initially grows quadratically (the so-called ballistic regime at $\gamma t' \ll 1$). In the limit of a long time (comparing to relaxation time $\gamma^{-1}$ rather than to force correlation time $\tau$ ), we have the diffusive growth $\langle (\Delta \mathbf{r})^2 \rangle \approx 6Tt'/M\gamma$.

Generally, $\langle (\Delta \mathbf{r})^2 \rangle = 6\kappa t$, where $\kappa$ is the diffusivity. If the particle radius $R$ is larger than the molecular mean free path $\ell$, in calculating resistance, we can consider liquid as a continuous medium and characterize it by the viscosity $\eta$. For a slow-moving particle, $v \ll v_T \ell/R$, the resistance is given by the Stokes formula, $\gamma = 6\pi \eta R/M$. The diffusivity then satisfies the Einstein relation:

$$\kappa = \frac{T}{M\gamma} = \frac{T}{6\pi \eta R}. \qquad (A.29)$$

The diffusivity depends on the particle radius but not the mass. Heavier particles are slower both to start and to stop moving. Measuring the diffusion of particles with a known size, one can determine the temperature.[2]

The probability distribution of displacement at $\gamma t' \gg 1$,

$$\rho(\Delta \mathbf{r}, t') = (4\pi \kappa t')^{-3/2} \exp[-|\Delta \mathbf{r}|^2/4\kappa t'], \qquad (A.30)$$

satisfies the diffusion equation $\partial \rho/\partial t' = \kappa \nabla^2 \rho$.

An external field $V(\mathbf{q})$ adds the force:

$$\dot{\mathbf{p}} = -\gamma \mathbf{p} + \mathbf{f} - \partial_q V, \quad \dot{\mathbf{q}} = \mathbf{p}/M. \qquad (A.31)$$

These equations characterize the system with the Hamiltonian $\mathcal{H} = p^2/2M + V(\mathbf{q})$. The system interacts with the thermostat, which provides friction $-\gamma \mathbf{p}$ and agitation $\mathbf{f}$—the balance between these two terms expressed by (A.27) means that the thermostat is in equilibrium.

We now pass from considering individual trajectories to the description of the "cloud" of trajectories and its statistics. Recall that our particle is macroscopic, that is, we consider the so-called overdamped limit $\gamma \tau \gg 1$, where $\tau$ is the correlation time of the random force. Since we are not interested in small, irregular changes of the velocity, but only in the statistics of displacement, we average (coarse-grain) over a moving time window, $p(t) \to p(t) = \int_{t-\tau}^{t+\tau} p(t')dt'$. After the average, we can neglect acceleration. In this limit, our

---

2. With the temperature in degrees, (A.29) contains the Boltzmann constant, $k = \kappa M\gamma/T$, which was actually determined by this relation and found truly constant, i.e., independent of the medium and the type of particle. That proved the reality of atoms—after all, $kT$ is the kinetic energy of a single atom.

second-order equation (A.31) on $\mathbf{q}$ is reduced to the first-order equation (we keep the same notations for coarse-grained quantities):

$$\gamma \mathbf{p} = \gamma M\dot{\mathbf{q}} = \mathbf{f} - \partial_q V. \tag{A.32}$$

We can now derive the equation on the probability distribution $\rho(\mathbf{q}, t)$, which changes with time due to random noise and evolution in the potential; the two mechanisms can be considered additively. Together, diffusion and advection give the Fokker-Planck equation, which is a multidimensional generalization of (4.7):

$$\frac{\partial \rho}{\partial t} = \kappa \nabla^2 \rho + \frac{1}{\gamma M} \frac{\partial}{\partial q_i} \rho \frac{\partial V}{\partial q_i} = -\operatorname{div} \mathbf{J}. \tag{A.33}$$

More formally, one can derive (A.33) by writing (A.32) as $\dot{q}_i - w_i = \eta_i$ and taking the random force Gaussian, delta correlated: $\langle \eta_i(0)\eta_j(t) \rangle = 2\kappa \delta_{ij}\delta(t)$. One can write the conditional probability $\rho(\mathbf{q}, t; 0, 0)$ as an average over all possible paths, each with its own weight determined by the Gaussian statistics of $\eta_i$. We start from the convolution identity, which simply states that the walker was certainly somewhere at an intermediate time $t_1$:

$$\rho(\mathbf{q}, t; 0, 0) = \int \rho(\mathbf{q}, t; \mathbf{q}_1, t_1)\rho(\mathbf{q}_1, t_1; 0, 0)\, d\mathbf{q}_1. \tag{A.34}$$

We now divide the time $t$ into a large number of short intervals, and using (A.30) for each interval, we write

$$\rho(\mathbf{q}, t; 0, 0) = \int \Pi_{i=0}^n \frac{d\mathbf{q}_{i+1}}{[4\pi \kappa (t_{i+1} - t_i)]^{d/2}}$$
$$\times \exp\left[ -\frac{|\mathbf{q}_{i+1} - \mathbf{q}_i + \mathbf{w}(t_{i+1} - t_i)|^2}{4\kappa (t_{i+1} - t_i)} \right]$$
$$\rightarrow \int \mathcal{D}\mathbf{q}(t') \exp\left[ -\frac{1}{4\kappa} \int_0^t dt' |\dot{\mathbf{q}} - \mathbf{w}|^2 \right]. \tag{A.35}$$

The last expression is an integral over paths that start at zero and end up at $\mathbf{q}$ at $t$. The notation $\mathcal{D}x(t')$ implies integration over the positions at intermediate times normalized by square roots of the time differences. The exponential gives the weight of every trajectory.

To describe the time change of $\rho$, consider the convolution identity (A.34) for an infinitesimal time shift $\epsilon$; then instead of the path integral, we get simply the integral over the initial position $\mathbf{q}'$. Into the exponent of this integral, we substitute $\dot{\mathbf{q}} = (\mathbf{q} - \mathbf{q}')/\epsilon$ and obtain

$$\rho(\mathbf{q}, t) = \int d\mathbf{q}' (4\pi\kappa\epsilon)^{-d/2} \exp\left[-\frac{[\mathbf{q} - \mathbf{q}' - \epsilon\mathbf{w}(\mathbf{q}')]^2}{4\kappa\epsilon}\right] \rho(\mathbf{q}', t - \epsilon).$$

$$(A.36)$$

What is written here is simply that the transition probability is the Gaussian probability of finding the noise $\eta$ with the right magnitude to provide for the transition from $\mathbf{q}'$ to $\mathbf{q}$. It is a coarse-grained continuous version of (4.3). We now change the integration variable, $\mathbf{y} = \mathbf{q}' + \epsilon\mathbf{w}(\mathbf{q}') - \mathbf{q}$, and keep only the first term in $\epsilon$: $d\mathbf{q}' = d\mathbf{y}[1 - \epsilon\partial_{\mathbf{q}} \cdot \mathbf{w}(\mathbf{q})]$. Here $\partial_{\mathbf{q}} \cdot \mathbf{w} = \partial_i w_i = div\,\mathbf{w}$. In the resulting expression, we expand the last factor, $\rho(\mathbf{q}', t - \epsilon)$:

$$\rho(\mathbf{q}, t) \approx (1 - \epsilon\partial_{\mathbf{q}} \cdot \mathbf{w}) \int d\mathbf{y}(4\pi\kappa\epsilon)^{-d/2} e^{-y^2/4\kappa\epsilon} \rho(\mathbf{q} + \mathbf{y} - \epsilon\mathbf{w}, t - \epsilon)$$

$$\approx (1 - \epsilon\partial_{\mathbf{q}} \cdot \mathbf{w}) \int d\mathbf{y}(4\pi\kappa\epsilon)^{-d/2} e^{-y^2/4\kappa\epsilon} \Big[\rho(\mathbf{q}, t) + (\mathbf{y} - \epsilon\mathbf{w}) \cdot \partial_{\mathbf{q}}\rho(\mathbf{q}, t)$$

$$+ (y_i y_j - 2\epsilon y_i w_j + \epsilon^2 w_i w_j)\partial_i\partial_j\rho(\mathbf{q}, t)/2 - \epsilon\partial_t\rho(\mathbf{q}, t)\Big]$$

$$= (1 - \epsilon\partial_{\mathbf{q}} \cdot \mathbf{w})[\rho - \epsilon\mathbf{w} \cdot \partial_{\mathbf{q}}\rho + \epsilon\kappa\,\Delta\rho - \epsilon\partial_t\rho + O(\epsilon^2)].$$

$$(A.37)$$

We obtain (A.33), collecting terms linear in $\epsilon$. Note that it was necessary to expand to the quadratic terms in $y$, which made the contribution linear in $\epsilon$ (namely the Laplacian, i.e., the diffusion operator).

## A.11    Fluctuation Relations in a Multidimensional Case

Apart from making the potential time-dependent (as in section 4.4), there is another way to deviate the system from equilibrium in more than one dimension: to add to (A.31) another external coordinate-dependent force $\mathbf{F}(\mathbf{q})$, which is nonpotential (not a gradient of any scalar):

$$\dot{\mathbf{p}} = -\gamma\mathbf{p} + \mathbf{f} - \partial_{\mathbf{q}}V + \mathbf{F}, \quad \dot{\mathbf{q}} = p/M.$$

The nonpotential force makes the system non-Hamiltonian even without any contact with a thermostat, that is, when $\gamma = 0$ and $\mathbf{f} = 0$. Bringing such a system into contact with a thermostat generally does not lead to thermal equilibrium, as we discussed in section 5.4. The equation on the full phase-space distribution $\rho(\mathbf{p}, \mathbf{q}, t)$ has the form

$$\partial_t \rho = \{\mathcal{H}, \rho\} + T\Delta_p \rho + \partial_\mathbf{p} \rho[\mathbf{F} - \gamma\mathbf{p}] = H_K \rho. \qquad (A.38)$$

It is called the Kramers equation. The Fokker-Planck equation follows from it in the overdamped limit. Only without $\mathbf{F}$, the Gibbs distribution $\exp(-\mathcal{H}/T)$ is a steady solution of (A.38), and one can formulate the detailed balance,

$$H_K^\dagger = \Pi e^{\mathcal{H}/T} H_K e^{-\mathcal{H}/T} \Pi^{-1}, \qquad (A.39)$$

where we add the operator-inverting momenta: $\Pi \mathbf{p} \Pi^{-1} = -\mathbf{p}$. A nonpotential force violates the detailed balance in the following way:

$$H_K^\dagger = \Pi e^{\mathcal{H}/T} H_K e^{-\mathcal{H}/T} \Pi^{-1} + (\mathbf{F} \cdot \dot{\mathbf{q}})/T. \qquad (A.40)$$

The last term (breaking the time-reversal symmetry) is again the power $(\mathbf{F} \cdot \dot{\mathbf{q}})$ divided by temperature, i.e., the entropy production rate. The work done by that force depends on the trajectory, in contrast to the case of a time-independent potential force. That dependence of the work on the trajectory precludes thermal equilibrium and is common for nonpotential forces and for time-dependent potential forces. A close analog of the Jarzynski relation can be formulated for the entropy production rate averaged during time $t$:

$$\sigma_t = \frac{1}{tT} \int_0^t (\mathbf{F} \cdot \dot{\mathbf{q}}) \, dt. \qquad (A.41)$$

The power $(\mathbf{F} \cdot \dot{\mathbf{q}})$ is identically zero for a magnetic Lorentz force, which is perpendicular to the velocity. For a potential force, $\mathbf{F} = dU/d\mathbf{q}$, we have $(\mathbf{F} \cdot \dot{\mathbf{q}}) = dU(\mathbf{q}(t))/dt$, and the integral turns into zero on average. A nonpotential external force $\mathbf{F}$ must on average do positive work to keep the system away from equilibrium. Over a long time, we thus expect $\sigma_t$ to be overwhelmingly positive, yet fluctuations do happen. The probabilities $P(\sigma_t)$ satisfy the relation, analogous to (4.20, A.23), which we give without derivation:

$$\frac{P(\sigma_t)}{P(-\sigma_t)} \propto e^{t\sigma_t} = e^{\Delta S}. \qquad (A.42)$$

The probability of observing a negative entropy production decays exponentially with the time of observation. Such fluctuations were unobservable in classical macroscopic thermodynamics, but they are often very important in modern applications to nano- and bio- objects. In the limit $t \to \infty$, the probability of the integral (A.41) must have a large-deviation form,

$P(\sigma_t) \propto \exp[-tH(\sigma_t)]$, so that (A.42) means that $H(\sigma_t) - H(-\sigma_t) = -\sigma_t$, as if $P(\sigma_t)$ is Gaussian with $H(\sigma_t) = (\sigma_t - 1)^2/2$.

One calls (4.20, A.23, A.42) detailed fluctuation-dissipation relations since they are stronger than integral relations of the type (4.17, 4.18). Indeed, it is straightforward to derive $\langle \exp(-t\sigma_t) \rangle = 1$ from (A.42).

The relation similar to (A.42) can be derived for any system symmetric with respect to some transformation, to which we add perturbation antisymmetric with respect to that transformation. Consider a system with the variables $s_1, \ldots, s_N$ and the even energy: $E_0(s) = E_0(-s)$. Consider the energy perturbed by an odd term, $E = E_0 - hM$, where $M(s) = \sum s_i = -M(-s)$. The probability of the perturbation $P[M(s)]$ satisfies the direct analog of (A.42), which is obtained by changing the integration variable $s \to -s$:

$$P(a) = \int ds \delta[M(s) - a] e^{\beta(ha - E_0)} = \int ds \delta[M(s) + a] e^{-\beta(ha + E_0)} = P(-a) e^{-2\beta ha}.$$

The validity condition for the results in this section, as well as in section 4.4, is that the interaction with the thermostat is represented by a noise independent of the evolution of the degrees of freedom under consideration.

## A.12 Quantum Fluctuations and Thermal Noise

Many aspects of the quantum world are bizarre and have no classical analog. In spite of that, quantum and thermal fluctuations impose uncertainty in somewhat similar ways due to the necessity of summing over different possibilities. One analogy is mentioned in section 4.4, where the Schrodinger equation is treated as a diffusion equation with an imaginary diffusivity. That means that one can treat the propagation of a quantum particle as a random walk in an imaginary time. Indeed, the transition probability of a classical unbiased random walk according to section A.10 is as follows:

$$\rho(\mathbf{x}, t; 0, 0) = \int \mathcal{D}\mathbf{x}(t') \exp\left[-\frac{1}{4\kappa} \int_0^t dt' \dot{\mathbf{x}}^2(t')\right].$$

The transition amplitude $T(\mathbf{x}, t; 0, 0)$ of a quantum nonrelativistic particle with mass $M$ from zero to $\mathbf{x}$ during $t$ is given by the sum over all possible paths connecting the points. Every path is weighted by the factor $\exp(iS/\hbar)$, where $S$ is the classical action (integral of energy over time):

$$T(\mathbf{x}, t; 0, 0) = \int \mathcal{D}\mathbf{x}(t') \exp\left[\frac{iM}{2\hbar} \int_0^t dt' \dot{\mathbf{x}}^2\right].$$

Comparing the two, we see that the transition probability of a random walk is given by the transition amplitude of a free quantum particle during an imaginary time.

Another similarity is revealed using the Heisenberg representation of the operators evolving in time. Recall that one special operator, called the Hamiltonian $\hat{\mathcal{H}}$, determines the temporal evolution of any other operator $\hat{P}$ according to $\hat{P}(t) = \exp(i\mathcal{H}t)P(0)\exp(-i\mathcal{H}t)$. We encountered an analog of the evolution operator $\hat{T}(t) = \exp(i\hat{\mathcal{H}}t)$ in section 4.3. The quantum-mechanical average of $\hat{P}(t)$ is calculated as a trace with the evolution operator normalized by the trace of the evolution operator:

$$\langle\hat{P}\rangle = \frac{\mathrm{Tr}\,\hat{T}(t)\hat{P}}{Z(t)}, \qquad Z(t) = \mathrm{Tr}\,\hat{T}(t) = \sum_a e^{-itE_a}. \qquad (A.43)$$

The normalization factor is naturally called the partition function, all the more so if we formally consider it for an imaginary time $t = i\beta$, now related to the inverse temperature:

$$Z(\beta) = \mathrm{Tr}\,\hat{T}(i\beta) = \sum_a e^{-\beta E_a}. \qquad (A.44)$$

If the inverse "temperature" $\beta$ goes to infinity, then all the sums are dominated by the ground state, $Z(\beta) \approx \exp(-\beta E_0)$, and the average in (A.44) is just the expectation value in the ground state.

That quantum-mechanical description can be compared with the so-called transfer-matrix description for the systems with nearest-neighbor interaction. Take for simplicity the Ising model whose Gibbs probability distribution, $\exp(-\beta\mathcal{H})$, is expressed via the classical Hamiltonian,

$$\mathcal{H} = \frac{J}{2}\sum_{i=1}^{N-1}(1-\sigma_i\sigma_{i+1}), \qquad \sigma_i = \pm 1. \qquad (A.45)$$

Consider it on a ring so that $\sigma_{N+1} = \sigma_1$ and write the partition function as a simple sum over a spin value at every site:

$$Z = \sum_{\{\sigma_i\}}\exp\left[-\frac{\beta J}{2}\sum_{i=1}^{N-1}(1-\sigma_i\sigma_{i+1})\right] \qquad (A.46)$$

$$= \sum_{\{\sigma_i\}}\prod_{i=1}^{N-1}\exp\left[-\frac{\beta J}{2}(1-\sigma_i\sigma_{i+1})\right]. \qquad (A.47)$$

Every factor in the product can have four values, which correspond to four different choices of $\sigma_i = \pm 1$, $\sigma_{i+1} = \pm 1$. Therefore, every factor can be written as an element of a $2 \times 2$ matrix: $\langle \sigma_j | \hat{T} | \sigma_{j+1} \rangle = T_{\sigma_j \sigma_{j+1}} = \exp[-\beta J(1 - \sigma_i \sigma_{i+1})/2]$. It is called a transfer matrix because it *transfers* us from one site to the next:

$$T = \begin{pmatrix} T_{1,1} & T_{1,-1} \\ T_{-1,1} & T_{-1,-1} \end{pmatrix}, \qquad (A.48)$$

where $T_{11} = T_{-1,-1} = 1$, $T_{-1,1} = T_{1,-1} = e^{-\beta J}$. For any matrices $\hat{A}, \hat{B}$, the matrix elements of the product are $[AB]_{ik} = A_{ij}B_{jk}$. Therefore, when we sum over the values of the intermediate spin, we obtain the elements of the matrix squared: $\sum_{\sigma_i} T_{\sigma_{i-1}\sigma_i} T_{\sigma_i \sigma_{i+1}} = [T^2]_{\sigma_{i-1}\sigma_{i+1}}$. The sum over $N-1$ spins gives $T^{N-1}$. Because of periodicity, we end up summing over a single spin, which corresponds to taking the trace of the matrix:

$$Z = \sum_{\{\sigma_i\}} T_{\sigma_1 \sigma_2} T_{\sigma_2 \sigma_3} \cdots T_{\sigma_N \sigma_1} = \sum_{\sigma_1 = \pm 1} \langle \sigma_1 | \hat{T}^{N-1} | \sigma_1 \rangle = \mathrm{Tr}\, T^{N-1}. \quad (A.49)$$

We thus see that taking the sum over two values of $\sigma$ at every site in the Ising model is the analog of taking a trace in the quantum-mechanical average. If there are $m$ values on the site, then $T$ is an $m \times m$ matrix. For a spin in $n$-dimensional space (described by the so-called $O(n)$ model), trace means integrating over orientations. The translations along the chain are analogous to quantum-mechanical translations in (imaginary) time. This analogy is not restricted to 1D systems; one can consider 2D strips that way too.

## A.13   Quantum Thermalization

At the end of section 6.3, we mentioned a purely quantum way of thermalization via the growth of entanglement. Is there any quantum analog of chaos that underlies thermalization the same way that dynamical chaos underlies mixing and thermalization in classical statistics, as described in section 5.3? Writing the classical formula of exponential separation, $\delta x(t) = \delta x(0)e^{\lambda t}$, as $\partial x(t)/\partial x(0) = e^{\lambda t}$ and replacing quantum mechanically the space derivative by the momentum operator, one naturally comes to consider the commutator of $\hat{x}(t)$ and $\hat{p}(0)$:

$$\frac{\partial x(t)}{\partial x(0)} = \frac{\partial x(t)}{\partial x(0)} \frac{\partial p(0)}{\partial p(0)} - \frac{\partial x(t)}{\partial p(0)} \frac{\partial p(0)}{\partial x(0)} = \{x(t), p(0)\} \rightarrow \hbar^{-1}[\hat{x}(t), \hat{p}(0)].$$

That corresponds to the Heisenberg representation, where operators are time-dependent. The commutator measures the effect of having the value $\hat{p}(0)$ on the later measurement of $\hat{x}(t)$. The average value of this commutator over the Gibbs distribution with a finite temperature $T$ is zero. Averaging the square, $C(t) = \langle([x(t)p(0)]^2)\rangle$, brings the concept of a so-called out-of-time-order correlation function like $\langle x(t)p(0)x(t)p(0)\rangle$. Such quantities are found to grow exponentially in time in some quantum systems (complicated enough to allow chaos and simple enough to allow for analytic solvability): $C(t) = \hbar^2 e^{2\lambda t}$, where the uncertainty relation gives the starting value at $t = 0$. The commutator squared is bounded, so that the exponential growth saturates when $C(t)$ becomes comparable with $\langle p^2\rangle\langle x^2\rangle$—that value is supposed to be much larger than $\hbar^2$, which requires a quasi-classical limit. The corresponding Lyapunov exponent dimensionally must be energy (temperature) divided by $\hbar$, and indeed $\lambda = 2\pi T/\hbar$ is shown to be a universal upper limit. To appreciate this, note that for a particle with mass $m$, the time of effective scattering $\lambda^{-1}$ cannot be less than the de Broglie wavelength $\hbar/\sqrt{mT}$ divided by the thermal velocity $\sqrt{T/m}$ (and the mass drops out!). The limit is reached, for instance, by black holes, which scramble quantum information at the greatest possible rate.

When there are many interacting particles, the growth of the many-particle version, $C_{ij} = \langle([x_i(t)p_j(0)]^2)\rangle$, describes how the entanglement of more and more distant particles appears on the way to thermalization. The evolution of the operators in the Heisenberg representation is governed by the Hamiltonian $\mathcal{H}\{\hat{x}_i, \hat{p}_i\}$:

$$\hat{x}_i(t) = e^{i\mathcal{H}t}\hat{x}_i(0)e^{-i\mathcal{H}t} = \sum_{j=0}^{\infty} \frac{(it)^j}{j!}[\mathcal{H}\ldots[\mathcal{H}\hat{x}_i(0)]\ldots]. \qquad (A.50)$$

Since the Hamiltonian describes the interaction between particles, the subsequent terms of the expansion involve more and more particles, which describes the growth of entanglement with time.

**Exercise A.4:** Growth of entanglement entropy.

Consider an Ising spin chain with the transverse magnetic field in the $x$ direction. The Hamiltonian $\mathcal{H} = \sum_j \sigma_i^z \sigma_{i+1}^z + h\sigma_i^x$. At $t = 0$, the chain is in a pure unentangled state, $\rho(0) = \rho_1 \otimes \rho_2 \otimes \ldots$, and all components, $\sigma_x, \sigma_y, \sigma_z$, are nonzero. Find in which order in $t$ entanglement between neighboring sites appears. Use the commutation relation $[\sigma_i^x, \sigma_j^z] = \hbar\delta_{ij}\sigma_i^y$.

## A.14 Exercises and Solutions

*1.1:* Candies and kids.

There are three candies and two systems to distribute them: system 1 contains two boys and system 2 contains three girls. Every boy and girl can have zero, one, two, or three candies with equal probability. Kids are distinguishable, but candies aren't. What is the most probable number of candies in system 1? What is the average number of candies in system 1? What are the most probable and average numbers of candies in system 2?

**Solution**

Let us compute the possible number of states. There is only one way to leave girls without candies so that $\Gamma_2(0) = 1$. Since we distinguish the boys, there are four ways to distribute the remaining three candies between the two of them: $(0, 3), (1, 2), (2, 1), (3, 0)$ so that $\Gamma_1(3) = 4$, and the composite system has $\Gamma_1(3)\Gamma_2(0) = 4 \cdot 1 = 4$ states. Similarly, $\Gamma_1(2)\Gamma_2(1) = 3 \cdot 3 = 9$, $\Gamma_1(1)\Gamma_2(2) = 2 \cdot 6 = 12$, $\Gamma_1(0)\Gamma_2(3) = 10$. Since all the states are equally probable, then the maximal probability corresponds to the maximal number of states $(12)$—when there is one candy in system 1 (and two in system 2). The probabilities of different numbers for system 1 are $p_1(0) = 10/(10 + 12 + 9 + 4) = 10/35$, $p_1(1) = 12/35$, $p_1(2) = 9/35$, $p_1(3) = 4/35$. The average number in system 1 is $(12 + 2 \cdot 9 + 3 \cdot 4)/35 = 42/35 = 6/5$. The average number in system 2 is $(9 + 2 \cdot 12 + 3 \cdot 10)/35 = 63/35 = 9/5$. The average number per kid is, thank God, the same for boys and girls: $3/5$. Most probable and average numbers approach each other only in the thermodynamic limit of large populations and plentiful candies.

*2.1:* Three squares have an average area of 100 $m^2$. The average of the lengths of their sides is 10 $m$. Use the Jensen inequality to determine the values the areas of the three squares can take.

**Solution**

Convexity of the parabola means that the only way to have a mean square equal to the square of the mean is to have all squares the same.

**2.2:** Information about precipitation.

In New York City, the probability of rain on the Fourth of July is 40%. On Thanksgiving, the probability of rain is 65%, while the probability of snow is 15%. When does the message on the presence or absence of precipitation bring more information—on Thanksgiving or on the Fourth of July?

### Solution

In the current climate, it is natural to assume that no snow is possible in July, so the information is $-0.4 \log_2 0.4 - 0.6 \log_2 0.6 \approx 0.97$ bits. On Thanksgiving, the probability of no precipitation is 0.2, so that the information is $-0.2 \log_2 0.2 - 0.8 \log_2 0.8 \approx 0.72$ bits.

**2.3:** Asking the right yes-no questions.

There are two different numbers not exceeding 100. What is the minimal number of one-bit questions we need to ask to determine both of them? How many bits does one need to establish $m$ numbers not exceeding $n$?

### Solution

There are $C_{100}^2 = 4950$ possible results, which all a priori have equal probability. The uncertainty is then $\log_2 4950 \approx 12.3$ bits. The minimal number of one-bit questions is 13. Each question needs to be designed to bring maximum information, which requires that the probabilities of "yes" and "no" are equal or as close as possible. One way to do that is to divide 4950 outcomes into two halves, etc. In a general case, the information we need is $\log_2 C_n^m$ bits.

**2.4:** Catching counterfeit coins.

In a pile of 27 coins, there is a counterfeit coin that weighs less than the others. What is the minimum number of weighings on a balancing scale we need to isolate that coin? Describe the procedure.

### Solution

The information we need is $\log_2 27$ bits. Every weighing has three possible outcomes—balanced, skewed to the left, skewed to the right—so it brings $\log_2 3$ bits. The number of weighings then cannot be less than $\log_2 27 / \log_2 3 = \log_3 27 = 3$. We get maximal information from every act if the probabilities of all three outcomes are equal. At the first step, that suggests dividing 27 into the three equal groups, and putting 9 coins on each side of the scale. That determines the suspicious 9 coins, which we then divide into three groups of 3, etc.

**2.5:** Deuteronomy.

Estimate the probability of the following sequence:

בראשית ברא אלוהים את השמים ואת הארץ

**Solution**

The length is $N = 29$ symbols. There are 12 distinct symbols with the number of appearances respectively 6, 3 (four symbols), 2 (four symbols), and 1 (three symbols). The total entropy is then $SN = 6\ln(29/6) + 12\ln(29/3) + 8\ln(29/2) + 3\ln(29) = 29\ln 29 - 18\ln 3 - 14\ln 2 \approx 68$. The probability is $\exp(-NS) = \exp(-68) \approx 3 \cdot 10^{-30}$. Without any extra information, it is an estimate from above, as any asymptotic equipartition. Assuming that this alphabet is approximately triple-redundant as any modern one, we would divide the entropy by 3.

**2.6:** Encoding by binary digits.

If we need to encode the results of independent throwing of a fair coin, we can use a one-bit encoding: 0 for heads and 1 for tails.

(a) If we have a fair die, which is either a regular tetrahedron or a cube, how long must our binary codewords be?
(b) If we have a fair die with six sides (all having the same probability), which binary encoding could we use to provide for a transmission rate within approximately 3% of the maximal rate?

**Solution**

(a) Two-bit encoding provides four words for the four sides of the tetrahedron, while three-bit encoding provides eight words for the eight sides of the cube. In both cases, those are the shortest encodings, so that the transfer rate is maximal.
(b) Each result brings the information $\log_2 6 \approx 2.58$ bits. We could divide the sequence of results into groups each containing $m$ results and encode each group by a word with length $n$; then we use $n/m$ bits per result. We can come within approximately 3% of 2.58 by using $n = 8$, $m = 3$, and $8/3 \approx 2.66$. Indeed, the

number of such groups is $6^3 = 216$. To represent them, we need eight-digit words, whose number is $2^8 = 256$.

**2.7:** Conditional entropy of criminality.

In our town, 2% of the people are criminals, and they all carry guns. In the rest of the population, only half of the people carry a gun.

(a) How much information yields a result about whether a given person is a criminal or not?

(b) How much information yields such a result if we also know in advance that the person does not carry a gun? How much information does the result yield if we see that the person carries a gun?

(c) How much information on average about a person's criminality yields knowledge of whether he/she carries a gun?

**Solution**

(a) Let us denote $B$ as the criminality status; then the probabilities are $p(B_1) = 0.02$ and $p(B_2) = 0.98$, so that the information about criminality yields
$$S(B) = -0.98 \log_2 0.98 - 0.01 \log_2 0.02 \approx 0.14 \text{ bits}.$$

(b) When we know that the person does not carry a gun (call it $A_1$), then the conditional probabilities are $p(B_1|A_1) = 0$ and $p(B_2|A_1) = 1$, so that the conditional entropy $S(B|A_1) = 0$; no uncertainty left. However, when we spot a gun (call it $A_2$), the conditional probabilities are $p(B_1|A_2) = 2/51$ and $p(B_2|A_2) = 49/51$, so that the conditional entropy
$$S(B|A_2) = -(49/51) \log_2 (49/51) - (2/51) \log_2 (2/51)$$
$\approx 0.24$ bits, which is larger that the unconditional $S(B)$. The reason is simple—seeing the gun has excluded 49% of the population, so that criminals now comprise a larger fraction, which makes probabilities closer and increases uncertainty.

(c) On average, the knowledge about carrying a gun decreases our uncertainty about criminality: $S(B|A) = p(A_1)S(B|A_1) + p(A_2)S(B|A_2) \approx 0.51 \cdot 0.24 \approx 0.12$ bits, which is less than $S(B) \approx 0.14$ bits. The information about criminality gained by the knowledge of gun carrying is $I(A, B) = S(B) - S(B|A) = 0.02$ bits. The same information about carrying a gun yields knowledge of criminality.

**2.8:** Cascade of binary channels.

Find the capacity of a cascade of $n$ consequent binary channels each with the probability of error $q$. How does the capacity decay at large $n$?

**Solution**

The capacity is $C = 1 - S(p_n)$, where $p_n$ is the probability of an error at the receiver after subsequently passing through $n$ channels. The error happens every time there is an odd number of errors on the way. For example, $p_3 = q^3 + 3q(1-q)^2$. For arbitrary $n$, the error probability can be found using binomials and making all the even terms cancel in the difference: $p_n = [(1-q+q)^n - (1-q-q)^n]/2 = [1 - (1-2q)^n]/2$. Asymptotically, the capacity is quadratic in $p_n - 1/2$ and decays exponentially with $n$: $C \propto (1-2q)^{2n}$.

**2.9:** Capacity of a noisy channel.

Consider a noisy channel $X \to Y$, where both input and output can take four values. After making 128 transmissions, the frequencies were as follows:

| $Y \backslash X$ | $x_1$ | $x_2$ | $x_3$ | $x_4$ | Sum |
|---|---|---|---|---|---|
| $y_1$ | 12 | 15 | 2 | 0 | 29 |
| $y_2$ | 4 | 21 | 10 | 0 | 35 |
| $y_3$ | 0 | 10 | 21 | 4 | 35 |
| $y_4$ | 0 | 2 | 15 | 12 | 29 |
| Sum | 16 | 48 | 48 | 16 | 128 |

Compute the mutual information between the input and the output. What fraction of the output $Y$ is a signal? What would be the capacity of the channel if it were error-free?

**Solution**

The table of joint probabilities $p(X, Y)$ looks as follows:

| $Y \backslash X$ | $x_1$ | $x_2$ | $x_3$ | $x_4$ | $p(Y)$ |
|---|---|---|---|---|---|
| $y_1$ | 0.094 | 0.117 | 0.016 | 0 | 0.227 |
| $y_2$ | 0.031 | 0.164 | 0.078 | 0 | 0.273 |
| $y_3$ | 0 | 0.078 | 0.164 | 0.031 | 0.273 |
| $y_4$ | 0 | 0.016 | 0.117 | 0.094 | 0.227 |
| $p(X)$ | 0.125 | 0.375 | 0.375 | 0.125 | 1 |

On the margins are the marginal probability distributions $p(X)$ and $p(Y)$.

It is straightforward now to compute $S(X) = 1.81$ bits, $S(Y) = 1.99$ bits, $S(X, Y) = 3.30$ bits. Note that the output entropy $S(Y)$ is larger than the input entropy $S(X)$, since the channel adds noise. The mutual information is as follows: $I(X, Y) = S(X) + S(Y) - S(X, Y) = 0.509$ bits. That means that only about a quarter of the output entropy is a signal; the rest is noise: $I(X, Y)/S(Y) = 0.509/1.99 \approx 0.256$. That value can be called transmission efficiency.

The error-free channel would have a capacity of two bits.

**2.10:** Efficient coding of Gaussian signals.

Consider two correlated signals with Gaussian statistics determined by $\langle x_1 \rangle = \langle x_2 \rangle = 0$, $\langle x_1^2 \rangle = \langle x_2^2 \rangle = 1$, and $\langle x_1 x_2 \rangle = r$. Find the most efficient encoding, $y_1(x_1, x_2)$ and $y_2(x_1, x_2)$. Remember that such encoding must maximize the data transmission rate, that is, the entropy.

**Solution**

We first apply a linear transform to statistically independent combinations: $x_+ = (x_1 + x_2)/\sqrt{2(1+r)}$, $x_- = (x_1 - x_2)/\sqrt{2(1 \pm r)}$, so that $\langle x_+^2 \rangle = \langle x_-^2 \rangle = 1$, $\langle x_+ x_- \rangle = 0$. Then we transform each of them to make the probability flat, using the *erf* function whose derivative is Gaussian:

$$y_{1,2} = erf\,(x_1 \pm x_2)/\sqrt{2(1-r)}\,.$$

**2.11:** Interaction information.

Consider a love triangle in which $Y$ can date $X$ and $Z$. Consider the statistics of dating-not dating. Compute the entropies of the joint distribution and all the marginal distributions and the interaction information, $II = S(X) + S(Y) + S(Z) + S(X, Y, Z) - S(X, Y) - S(X, Z) - S(Y, Z)$, in the two cases.

(a) Assume that $Y$ with equal $1/3$ probabilities can be in these three states: not dating anyone, dating $X$, dating $Z$. That is, $Y$ is dating with probability $2/3$.

(b) Assume that $Y$ with equal $1/4$ probabilities can be in these four states: not dating anyone, dating $X$, dating $Z$, dating both $X$ and $Z$.

**Solution**

(a) Both $X$ and $Z$ are dating with probability $1/3$ and not dating with probability $2/3$. $Y$ is dating with probability $2/3$, so that

$S(X) = S(Z) = S(Y) = (1/3) \log_2 3 + (2/3) \log_2 (3/2)$. All the composite states have probability $1/3$ so that $S(X, Y) = S(Y, Z) = S(X, Z) = S(X, Y, Z) = \log_2 3$. In this case, $II < 0$. Even though there is a true correlation between the dating states of $X$ and $Z$, knowledge of whether $Y$ is dating or not increases it.

(b) $S(X) = S(Z) = 1, S(Y) = 1/2 + (3/4) \log_2 (4/3), S(X, Y, Z) = S(X, Z) = 2, S(X, Y) = S(Z, Y) = 3/2. \, II = 1 + 1 + 1/2 + (3/4) \log_2 (4/3) + 2 - 2 - 3 = (1/4) \log_2 (16/27) < 0$. In this case, $I(X, Z) = 0$ and $II = -I(X, Z|Y)$; that is, knowledge of the dating state of $Y$ imposes a correlation between $X$ and $Z$. Alternatively, one may say that this is a case of synergy, where knowing the dating states of both $X$ and $Z$ gives more information about $Y$ than knowing them separately.


**2.12:** Correlations between three events.

What sign is the interaction information between i) clouds, rain, and darkness, and ii) a dead car battery, a broken fuel pump, and failure to start the engine?

**Solution**

i) Positive interaction information appears for common-cause phenomena. Clouds cause rain and also block the sun; therefore, the correlation between rain and darkness is partly accounted for by the presence of clouds: $I(\text{rain}, \text{dark}|\text{cloud}) < I(\text{rain}, \text{dark})$ and $II(\text{rain}, \text{dark}, \text{cloud}) > 0$.

ii) Negative interaction information appears for common-effect phenomena. Generally, malfunctions of the fuel pump and the battery are uncorrelated. However, if we know that the engine failed to start (fix common effect), then the dependency appears: if the check shows that the battery is OK, we infer that the pump must be broken.


**2.13:** Distance between distributions.

Consider two random quantities $X$ and $Y$ and define $\rho(X, Y) = S(X|Y) + S(Y|X)$. Apparently, $\rho(X, Y)$ is nonnegative and turns into zero if and only if $X$ and $Y$ are perfectly correlated.

(a) Prove the triangle inequality $\rho(X, Z) \leq \rho(X, Y) + \rho(Y, Z)$.
(b) Recall that the three random quantities $X \to Y \to Z$ constitute a Markov triplet if $Y$ is completely determined by $X, Z$, while $X, Z$ are independent, conditional on $Y$; that is, $I(X, Z|Y) = 0$. Find the relation between $\rho(X, Z)$ and $\rho(X, Y)$, $\rho(Y, Z)$.

**Solution**

(a) Triangle inequality:

$$S(X, Z) \leq S(X, Y, Z) = S(X, Z|Y) + S(Y)$$

$$\leq S(X|Y) + S(Z|Y) + S(Y) = S(X, Y) + S(Y, Z) - S(Y).$$

(b) For the Markov triplet, $Y$ is completely determined by $X, Z$, that is, $S(X, Y, Z) = S(X, Z)$. In addition, $I(X, Z|Y) = S(X, Z|Y) - S(X|Y) - S(Z|Y) = S(X, Y, Z) + S(Y) - S(X, Y) - S(Y, Z) = 0$. That gives equality $\rho(X, Z) = \rho(X, Y) + \rho(Y, Z)$, that is, the "point" $Y$ lies on a "straight line" through $X$ and $Z$. In terms of the distances, all points in a Markov chain lie in a straight line. That allows one to establish identities of the following type: $S(X_1, X_3) + S(X_2, X_4) = S(X_1, X_4) + S(X_2, X_3)$.

**3.1:** Distribution from information.

Consider particles having coordinates $x$ on a line: $-\infty < x < \infty$. Find the probability distribution $p(x)$ in two cases.

(a) The only information established by measurement is that the mean distance from zero is $\langle |x| \rangle = X$.
(b) The only information established by measurement is that the variance is given by $\langle x^2 \rangle = X^2$.

Which measurement provides more information on the coordinate distribution? Quantify the difference in bits.

**Solution**

$p_1 = (2X)^{-1} \exp(-|x|/X)$, $p_2 = (2\pi X^2)^{-1/2} \exp(-x^2/2X^2)$. $S_1 = \ln(2X) + 1$, $S_2 = \ln X + 1/2[1 + \ln(2\pi)]$. $I_2 - I_1 = (S_1 - S_2)/\ln 2 = (1 + \ln 2 - \ln \pi)/2 \ln 2 \approx 0.4$ bits.

**3.2:** Rate-distortion function of a binary source.

Consider a binary source, which generates $B = 1$ with probability $p < 1/2$ and $B = 0$ with probability $1 - p$. Define the distortion function $d(A, B) = \delta_{AB} - 1$, that is, zero when $A = B$ and unity otherwise (the so-called Hamming function). Find the rate-distortion function $R(D)$.

### Solution

For $D > p$, we can choose $A = 0$ with probability 1, so that $R = I(A, B) = 0$. For $D \leq p$, one realizes the minimum of the mutual information by symmetric encoding: $P(A = 1|B = 1) = 1 - D = P(A = 0|B = 0)$ and $P(A = 1|B = 0) = P(A = 0|B = 1) = D$. By the nature of the Hamming distortion function, $D$ is the probability of miscoding. The conditional probabilities give $S(B|A) = S(D)$ and $R(D) = I(A, B) = S(B) - S(B|A) = S(p) - S(D)$, where $S(x) = -x \log x - (1 - x) \log(1 - x)$.

**3.3:** Bookmaker's sure bet.

In a series of two-horse races, the first horse wins three times more often than the second one. Yet public sentiment is such that it bets on the first horse only twice as many times. A bookmaker has two choices to set the rewards: i) according to race probabilities, pay respectively $4/3Z$ and $4/Z$ times the amount of the bet on the first/second horse, ii) according to public preferences, pay respectively $3/2Z$ and $3/Z$ times the amount of the bet on the first/second horse. Here $Z > 1$ to guarantee a (small) profit. Which strategy is preferable?

### Solution

Even though the longtime mean profit is the same in both cases, the second strategy is preferable because the bookmaker guarantees the profit of $1 - 1/Z$ from every race independent of its result. The reason is that every bet is an average over many people. The first strategy can encounter some short-term losses before the long-term average over many races sets it.

**4.1:** PageRank of the two-page internet.

Consider the simplest version of the internet, which has two pages: page 1 has one link to page 2, which has no links. Rank these pages according to the PageRank algorithm with arbitrary $d < 1$ (the probability of following a link).

**Solution**

The Google matrix is as follows:

$$\hat{G} = \frac{1}{2} \begin{pmatrix} 1 & 1 \\ 1-d & 1+d \end{pmatrix}.$$

The eigenvalues satisfy the equation $\|\lambda \hat{I} - \hat{G}\| = (\lambda - 1/2)[\lambda - (1+d)/2] - (1-d)/4 = 0$, which has two solutions: $\lambda = 1$ and $\lambda = d/2$. The eigenvector that corresponds to $\lambda = 1$ is $(1 - d, 1)/(2 - d)$, so the rank 1 is $(1 - d)/(2 - d)$ and the rank 2 is $1/(2 - d)$.

**4.2:** Eigenvalues of the Google matrix.

Assume that the matrix $\hat{A}$ with the spectrum $(1, \lambda_2, \dots, \lambda_n)$ is stochastic, that is, $\sum_j a_{ij} = 1$ for every $i$. Prove that the spectrum of the Google matrix $\hat{G} = d\hat{A} + (1 - d)\mathbf{e}\mathbf{v}^T$ is $(1, d\lambda_2, \dots, d\lambda_n)$, where $\mathbf{v}$ is an arbitrary probability vector, that is, $\sum_i v_i = 1$.

**Solution**

Any stochastic matrix has the eigenvector $\mathbf{e} = (1, \dots, 1)$ with the eigenvalue 1. Let $\hat{Q} = (\mathbf{e}\ \hat{X})$ be a nonsingular matrix having the eigenvector $\mathbf{e}$ as its first column. Denote $\hat{Q}^{-1} = \begin{pmatrix} \mathbf{y}^T \\ \mathbf{Y}^T \end{pmatrix}$. Then $\hat{Q}^{-1}\hat{Q} = \begin{pmatrix} \mathbf{y}^T\mathbf{e} & \mathbf{y}^T\hat{X} \\ \mathbf{Y}^T\mathbf{e} & \mathbf{Y}^T\hat{X} \end{pmatrix}$, which gives two useful identities: $\mathbf{y}^T\mathbf{e} = 1$ and $\mathbf{Y}^T\mathbf{e} = 0$. Now we can use $\hat{Q}$ in the similarity transformation (which does not change the spectrum) to isolate the first eigenvalue:

$$\hat{Q}^{-1}\hat{A}\hat{Q} = \begin{pmatrix} \mathbf{y}^T\mathbf{e} & \mathbf{y}^T\hat{A}\hat{X} \\ \mathbf{Y}^T\mathbf{e} & \mathbf{Y}^T\hat{A}\hat{X} \end{pmatrix} = \begin{pmatrix} 1 & \mathbf{y}^T\hat{A}\hat{X} \\ 0 & \mathbf{Y}^T\hat{A}\hat{X} \end{pmatrix}, \qquad (A.51)$$

which reveals that $\mathbf{Y}^T\hat{A}\hat{X}$ contains the remaining eigenvalues of $\hat{A}$, $(\lambda_2, \dots, \lambda_n)$. Applying the similarity transformation to $\hat{G} = d\hat{A} + (1 - d)\mathbf{e}\mathbf{v}^T$ gives

$$\hat{Q}^{-1}\hat{G}\hat{Q} = \begin{pmatrix} d & d\mathbf{y}^T\hat{A}\hat{X} \\ 0 & d\mathbf{Y}^T\hat{A}\hat{X} \end{pmatrix} + (1 - d)\begin{pmatrix} \mathbf{y}^T\mathbf{e} \\ \mathbf{Y}^T\mathbf{e} \end{pmatrix}(\mathbf{v}^T\mathbf{e}\ \mathbf{v}^T\hat{X})$$

$$= \begin{pmatrix} 1 & d\mathbf{y}^T\hat{A}\hat{X} + (1 - d)\mathbf{v}^T\hat{X} \\ 0 & d\mathbf{Y}^T\hat{A}\hat{X} \end{pmatrix}.$$

Comparing with (A.51), we conclude that the eigenvalues of $\hat{G} = d\hat{A} + (1 - d)\mathbf{ev}^T$ are $(1, d\lambda_2, \ldots, d\lambda_n)$.

**4.3: Solus Rex.**

A king randomly moves to any of the adjacent squares with equal probability on an otherwise empty $3 \times 3$ chessboard.

(a) How much information brings a message specifying his position?
(b) If we wish to encode the whole long game (the random walk of the king), we need to know how the number of typical sequences $N(n)$ grows asymptotically with the number $n$ of the moves: $\lim_{n\to\infty} N(n) = 2^{nS}$. Find $S$, which is called the information rate of the source. Is it the same as the entropy that determined the answer to the previous question?

**Solution**

Denote

$$\begin{pmatrix} 1 & 2 & 3 \\ 4 & 5 & 6 \\ 7 & 8 & 9 \end{pmatrix}.$$

(a) A message specifying the king's position gives the information equal to the entropy of the invariant (stationary) distribution $p_i, i = 1, \ldots, 9$. A straightforward way to find the distribution is to write the $9 \times 9$ transition probability matrix $\hat{A} = \{a_{ij}\}$ and find its eigenvector with the unit eigenvalue:

$$p_i = \sum_j p_j a_{ji}. \tag{A.52}$$

A simpler way is to use symmetries and a detailed balance. Symmetries require $p_1 = p_3 = p_7 = p_9, p_2 = p_4 = p_6 = p_8$, and normalization gives $4p_1 + 4p_2 + p_5 = 1$. The detailed balance means that for any $i, j$ we have

$$p_i a_{ij} = p_j a_{ji}. \tag{A.53}$$

Note that (A.52) is a sum of (A.53) over $j$. The detailed balance gives two independent relations, $p_1/3 = p_2/5$ and $p_1/3 = p_5/8$. That gives $p_1 = p_3 = p_7 = p_9 = 3/40, p_2 = p_4 = p_6 = p_8 = 1/8$,

$p_5 = 1/5$. The entropy of such a distribution is $S = -\sum p_i$ $\log_2 p_i = 29/10 + (1/2) \log_2 5 - (3/10) \log_2 3$.

(b) The information rate is $H = -\sum_{ij} p_i a_{ij} \log_2 a_{ij} = 3/40 + 10^{-1} \log_2 15$. Of course, $H < S$ since the rules of the game impose restrictions on possible strings.

### 4.4: Random walk on a circle.

Consider a one-dimensional random walk over a circle with $N$ sites as a Markov chain and write the one-step transformation of the probability distribution over the sites $i = 1, \ldots, N$. Find the transition probability matrix $\hat{A}$ and show that its eigenvectors are $e^{ijk}$ if $k_n = 2\pi n/N$ for $n = 0, 1, \ldots, N - 1$. Show that the only stationary distribution is the eigenvector with the highest eigenvalue and the rate of relaxation to it is determined by the second largest eigenvalue.

#### Solution

Probability at a site evolves according to $p_i(t + \tau) = [p_{i-1}(t) + p_{i+1}(t)]/2 = \sum_j p_j a_{ij}$; that is, the matrix $\hat{A}$ has all elements zero, except next-to-diagonal elements $1/2$. Eigenvalues are $\lambda_n = \cos k_n$ for the respective eigenvectors $e^{ijk_n}$. After $m$ steps in time, the amplitude corresponding to an eigenvector is multiplied by $\lambda_n^m \approx \exp[-2m(n\pi/N)^2]$. In the limit $m \to \infty$, the only surviving eigenvector corresponds to $\lambda = 1$ and $n = 0$; the longtime relaxation is determined by the next eigenvector with $n = 1$. The relaxation time is then $\tau N^2/2\pi^2$.

### 4.5: Random walk in an inverted potential.

Consider a particle in an inverted quadratic potential $V(x) = -\alpha x^2/2$ under the action of a random noise $\eta(t)$ with $\langle \eta(0)\eta(t) \rangle = \delta(t)$. This is described by the Langevin equation with $\alpha > 0$:

$$\dot{x} = \alpha x + \eta. \tag{A.54}$$

Assume that the particle is at $x_0$ at $t = 0$.

(a) Find the probability distribution $\rho(x, t)$ by directly solving (A.54). Find the longtime decay of probability at a finite distance.

(b) Write the Fokker-Planck Hamiltonian $H_{FP}$. Find the spectrum of the Hamiltonian and compare it with the cases of negative and zero $\alpha$. In our case of positive $\alpha$, relate the longtime asymptotic of $\rho(x, t)$ to the lowest eigenvalue of the Fokker-Planck Hamiltonian.

## Solution

(a) At all times, $x$ has Gaussian statistics since it is linearly related to $\eta$. To specify the Gaussian statistics, all we need is the mean and the variance. The direct solution of the Langevin equation gives

$$x(t) = x_0 e^{\alpha t} + \int_0^t e^{\alpha(t-t')} \eta(t') \, dt' . \tag{A.55}$$

The mean value grows exponentially with time: $\langle x(t) \rangle = x_0 e^{\alpha t}$. Taking the square of (A.55), we obtain the variance:

$$\langle [x(t) - \langle x(t) \rangle]^2 \rangle = \frac{1}{2\alpha} \left( e^{2\alpha t} - 1 \right). \tag{A.56}$$

The distribution is thus given by

$$\rho(x, t) = \left[ \frac{\pi}{\alpha} \left( e^{2\alpha t} - 1 \right) \right]^{-1/2} \exp \left[ -\frac{\alpha}{e^{2\alpha t} - 1} \left( x - x_0 e^{\alpha t} \right)^2 \right]. \tag{A.57}$$

Asymptotically at $\alpha t \gg 1$, the probability at any finite distance, $x \ll x_0 e^{\alpha t}$, decays as follows:

$$\rho(x, t) = \left( \frac{\alpha}{\pi} \right)^{1/2} \exp\left( -\alpha t - \alpha x_0^2 \right).$$

(b) The Fokker-Planck Hamiltonian is $H_{FP} = \left( -\partial_x^2 + \alpha^2 x^2 + \alpha \right)/2$. Its eigenvalues are $E_n = (n + 1/2)|\alpha| + \alpha/2$. For a noninverted potential with $\alpha < 0$, the lowest eigenvalue is zero, $E_0 = 0$, and the respective eigenfunction is an asymptotic equilibrium steady state, $\exp(\alpha x^2)$. In our case, however, the lowest eigenvalue is positive: $E_0 = \alpha$. Still, it determines the longtime asymptotic of (A.57): $\rho(x, t) \propto e^{-\alpha t} + O\left(e^{-2\alpha t}\right)$. The probability of finding the particle at any finite distance from the origin decays exponentially with the rate, $E_0 = \alpha$. In the degenerate case of a free random walk, $\alpha = 0$, the spectrum of $H_{FP}$ is continuous and covers the half-line $[0, \infty)$; the probability decays by a power law, $t^{-1/2}$.

**5.1:** RG and the family of universal distributions.

Consider a set of random iid variables $x_1 \dots x_N$.

(a) The RG reduces the number of random variables by replacing any two of them by their mean (half sum): $z_i = (x_{2i-1} + x_{2i})/2$. Show that the Fourier image of the distribution $\rho(k) = exp(-|k|)$ is a fixed point of this map. Study the linear stability of this fixed point. What probability density does this correspond to? Why doesn't this contradict the central limit theorem?

(b) Consider the one-parametric family of the transformations: $z_i = (x_{2i-1} + x_{2i})/2^{1/\mu}$. Find the fixed point, that is, the distribution invariant under this transformation.

**Solution**

(a) The RG transformation is as follows:

$$\rho'(z) = \int \rho(x)\rho(y)\delta(z - x/2 - y/2)\, dxdy \Rightarrow \rho'(k)$$

$$= \rho^2(k/2).$$

The stationary solution satisfying $\rho_0(k) = \rho_0^2(k/2)$ is $\rho_0(k) = exp(-|k|)$, whose Fourier transform is $\rho_0(x) = (1 + x^2)^{-1}$. For $\rho(k) = \rho_0(k)[1 + h(k)]$, the linear eigenfunctions of the RG transformation are $h_m = k^m$ with eigenvalues $2^{1-m}$. The transform conserves normalization and the mean ($m = 0, 1$), so $\rho_0$ is the limiting distribution (attractor) when the mean was zero from the very beginning. Since every step of the transformation diminishes the variance, then the invariant distribution $\rho_0(x)$ must have an infinite variance. This is the reason the central limit theorem cannot be applied. In a sense, this distribution (named after Cauchy) generalizes the central limit theorem for iid variables with the tails $1/x^2$; it can be generalized further for other power tails.

When the second moment is finite, all terms within the sum are of the same order, and none of them play a dominant role, which leads to a sum having Gaussian statistics. When the second moment is infinite, a few terms become extremely large and dominate the sum, which leads to a non-Gaussian distribution.

(b) We now have the family of universal distributions $\rho(k) = \exp(-|k|^\mu)$, characterized by the parameter $\mu$.

***6.1:*** Least uncertain wave packet.

Proceeding from the fact that the momentum operator in the coordinate representation is $\hat{p}_x = \imath\hbar\partial_x$, find the state $\psi(x)$ that minimizes the expectation of the product of variances of the coordinate and momentum. What is the corresponding $\psi(p)$?

**Solution**

Consider (6.1):

$$|\langle\psi|[\hat{A},\hat{B}]|\psi\rangle|^2 = 4|\langle\psi|\hat{A}\hat{B}|\psi\rangle|^2 - |\langle\psi|\hat{A}\hat{B}+\hat{B}\hat{A}|\psi\rangle|^2$$

$$\leq 4|\langle\psi|\hat{A}\hat{B}|\psi\rangle|^2 \leq 4\langle\psi|\hat{A}^2|\psi\rangle\langle\psi|\hat{B}^2|\psi\rangle.$$

To turn the Cauchy-Schwarz inequality into equality in the second line, we need the vectors $\hat{A}|\psi\rangle$ and $\hat{B}|\psi\rangle$ to be parallel. Solving the equation

$$\hat{x}|\psi\rangle = x\psi(x) \propto \hat{p}|\psi\rangle = \imath\hbar\frac{\partial\psi(x)}{\partial x},$$

we obtain $\psi(x) = \exp(\imath Cx^2)$. In addition, we need $\langle\psi|\hat{p}\hat{x}+\hat{x}\hat{p}|\psi\rangle = 0$, which requires imaginary $C = \imath a$. Normalization implies $a > 0$. The solution is thus Gaussian: $\psi(x) \propto \exp(-ax^2)$. The wave functions of a quantum particle in the coordinate and momentum representations are related by the Fourier transform, $\psi(x) = \int dp\psi(p)\exp(\imath px/\hbar)$, so that $\psi(p)$ is Gaussian as well. Computing the respective probability distributions $|\psi(x)|^2$ and $|\psi(p)|^2$, one verifies that the Heisenberg minimum of the product of variances of $x$ and $p$ is realized by a Gaussian wave packet.

***6.2:*** Density matrix.

Consider two mixed states (ensembles): In the first ensemble $A$, the system can be in the state $|0\rangle$ with the probability $3/4$ and in the state $|1\rangle$ with the probability $1/4$. In the second ensemble $B$, the system can be in the state $|a\rangle = \sqrt{3/4}\,|0\rangle + \sqrt{1/4}\,|1\rangle$ and in the state $|b\rangle = \sqrt{3/4}\,|1\rangle - \sqrt{1/4}\,|0\rangle$ with equal probability.

(a) Write the density matrices for these two ensembles in the basis $|0\rangle, |1\rangle$.

(b)  Consider two sets of normalized vectors, $|\psi_i\rangle$ and $|\phi_j\rangle$, and two probability distributions, $p_i$ and $q_j$. The sets are related by $\sqrt{p_i}\,|\psi_i\rangle = \sum_j u_{ij}\sqrt{q_j}\,|\phi_j\rangle$, where the matrix is unitary: $\sum_i u_{ij}u_{ik}^* = \delta_{jk}$. Find the relation between two density matrices, $\rho_1 = \sum_i p_i\,|\psi_i\rangle\,\langle\psi_i|$ and $\rho_2 = \sum_j q_j\,|\phi_j\rangle\,\langle\phi_j|$.

**Solution**

(a)  The density matrices are the same:

$$\rho_A = \rho_B = \frac{3}{4}\,|0\rangle\,\langle 0| + \frac{1}{4}\,|1\rangle\,\langle 1| = \frac{1}{4}\begin{bmatrix} 3 & 0 \\ 0 & 1 \end{bmatrix}.$$

(b)  More generally,

$$\rho_1 = \sum_i p_i\,|\psi_i\rangle\,\langle\psi_i| = \sum_{ijk} u_{ij}u_{ik}^*\sqrt{q_j q_k}\,|\phi_j\rangle\,\langle\phi_k|$$

$$= \sum_{jk} \delta_{jk}\sqrt{q_j q_k}\,|\phi_j\rangle\,\langle\phi_k| = \sum_j q_j\,|\phi_j\rangle\,\langle\phi_j| = \rho_2.$$

**6.3:** Von Neumann entropy.

Consider two nonorthogonal states, $|0\rangle$ and the superposition $|s\rangle = (|0\rangle - |1\rangle)/\sqrt{2}$, mixed with the respective probabilities $p$ and $1-p$. Find the density matrix $\rho$ in the orthogonal basis $|0\rangle$, $|1\rangle$, diagonalize it, compute the von Neumann entropy $S(\rho)$, and compare it with $S(p)$.

**Solution**

Since the probabilities of the nonorthogonal states $|0\rangle$ and $(|0\rangle + |1\rangle)/\sqrt{2}$ are, respectively, $p$ and $1-p$, the density matrix in the orthogonal basis $|0\rangle$, $|1\rangle$ is as follows:

$$\rho = p\,|0\rangle\,\langle 0| + \frac{1-p}{2}\,|0+1\rangle\,\langle 0+1| = \frac{1}{2}\begin{bmatrix} 1+p & 1-p \\ 1-p & 1-p \end{bmatrix}. \qquad \text{(A.58)}$$

The eigenvalues are $q = (1 \pm \sqrt{2p^2 - 2p + 1})/2$ and $1 - q$ with the respective eigenstates $\sqrt{q}\,|1\rangle + \sqrt{1-q}\,|0\rangle$ and $-\sqrt{q}\,|0\rangle + \sqrt{1-q}\,|1\rangle$. The von Neumann entropy is the Shannon entropy in this orthonormal representation: $S(\rho) = S(q)$. One can show that $S(q) \leq S(p)$ for any $p$.

In particular, for $p = 1/2$, we have $q = (2 + \sqrt{2})/2 = \sin^2(\pi/8)$ and

$$S(\rho) = S(q) = -q \log q - (1 - q) \log(1 - q)$$

$$= 1 + \frac{1}{2} - \frac{1}{2\sqrt{2}} \log \frac{2 + \sqrt{2}}{2 - \sqrt{2}} \approx 0.6 \text{ bits,}$$

which is indeed less than $S(p) = 1$.

**A.1:** Large deviations for the energy of particles.

Find the probability distribution of the kinetic energy, $E = \sum_1^N p_i^2/2$, of $N$ classical, identical unit-mass particles in 1D, which have the Maxwell distribution over momenta. Derive the large-deviation form of the distribution in the limit $N \to \infty$.

**Solution**

The Maxwell distribution for momenta is Gaussian:

$$\rho(p_1, \ldots, p_N) = (2\pi T)^{-N/2} \exp\left(-\sum_1^N p_i^2/2T\right).$$

The energy probability for any $N$ is done by integration, using spherical coordinates in the momentum space:

$$\rho(E, N) = \int \rho(p_1, \ldots, p_N)\delta\left(E - \sum_1^N p_i^2/2\right) dp_1 \ldots dp_N$$

$$= \left(\frac{E}{T}\right)^{N/2} \frac{\exp(-E/T)}{E\Gamma(N/2)}. \tag{A.59}$$

Plotting it for different $N$, one can appreciate how the thermodynamic limit appears. Taking the logarithm and using the Stirling formula, one gets the large-deviation form for the energy $R = E/\bar{E}$, normalized by the mean energy $\bar{E} = NT/2$:

$$\ln \rho(E, N) = \frac{N}{2} \ln \frac{RN}{2} - \ln \frac{N}{2}! - \frac{RN}{2} \approx \frac{N}{2}(1 - R + \ln R). \tag{A.60}$$

This expression has a maximum at $R = 1$, i.e., the most probable value is the mean energy. The probability of $R$ is Gaussian near the maximum when $R - 1 \leq N^{-1/2}$ and non-Gaussian for larger deviations. Notice that this function is not symmetric with respect to the minimum; it is logarithmic at zero and linear asymptotic at infinity.

**A.2:** Large deviations for binomial distribution.

One of the most widely used statistical distributions (including in this book) is the binomial distribution of two possible outcomes, $y = 1$ with probability $p$ and $y = 0$ with probability $1 - p$. Compute the probability that, in a large number of trials $N$, the first outcome happens $qN$ times; that is, $X = \sum_{i=1}^{N} y_i = qN$. Do it in two ways: 1) discrete combinatoric, using the binomial formula $C_N^{qN} = N!/(qN)!(N - qN)!$ for the number of ways to choose $qN$ out of $N$, and the Stirling formula $\ln N! \approx N \ln N$; 2) continuous, using the large-deviation theory, that is, computing the cumulant generating function $G(z) = \ln\langle e^{zy} \rangle = \ln(pe^z + 1 - p)$ and the Legendre transform of it.

**Solution**

1) $\mathcal{P}(q) = C_N^{qN} p^{qN} (1 - p)^{(1-q)N}$ and $\ln[\mathcal{P}(q)] \approx N \ln N - qN \ln(qN) - (1 - q)N \ln(N - qN) + qN \ln p + (1 - q)N \ln(1 - p) = qN \ln(q/p) + (1 - q)N \ln[(1 - q)/(1 - p)]$.

2) $\mathcal{P}(q) \approx \exp[-NH(q)]$, where $H(q) = z_0 q - G(z_0)$, where $G'(z_0) = q$. Inverting $G'(z_0)$, we obtain $z_0 = \ln[q(1 - p)/p(1 - q)]$, so that $H(q) = q \ln(q/p) + (1 - q) \ln[(1 - q)/(1 - p)]$. It is a concave function with the minimum $H(q = p) = 0$.

We see that $H$ is the relative entropy between the measured probability $q$ and the probability $p$ treated as a hypothesis.

**A.3:** Generating function for cumulants.

The derivatives at zero of the logarithm of the generating function $G(z) = \ln\langle e^{zy} \rangle$ are called cumulants. Are $\kappa_n = (d^n G/dz^n)_{z=0}$ equal to the moments of $(y - \langle y \rangle)^n$? Express the first four $\kappa_1, \ldots, \kappa_4$ via $\mu_n = \langle y^n \rangle$.

**Solution**

$$\langle \exp(zy) \rangle = 1 + \sum_{n=1}^{\infty} \frac{z^n}{n!} \langle y^n \rangle, \quad G(z) = \ln\langle e^{zy} \rangle = \ln\langle 1 + e^{zy} - 1 \rangle$$

$$= -\sum_{n=1}^{\infty} \frac{1}{n} (1 - \langle \exp(zy) \rangle)^n = -\sum_{n=1}^{\infty} \frac{1}{n} \left( -\sum_{m=1}^{\infty} \frac{z^m}{m!} \langle y^m \rangle \right)^n$$

$$= z\langle y\rangle + \left(\langle y^2\rangle - \langle y\rangle^2\right)\frac{z^2}{2!} + \dots \tag{A.61}$$

Comparing order by order, we find

$$\kappa_1 = \mu_1, \; \kappa_2 = \mu_2 - \mu_1^2 = \langle(y - \mu_1)^2\rangle,$$

$$\kappa_3 = \mu_3 - 3\mu_2\mu_1 + 2\mu_1^3 = \langle(y - \mu_1)^3\rangle,$$

$$\kappa_4 = \mu_4 - 4\mu_3\mu_1 - 3\mu_2^2 + 12\mu_2\mu_1^2 - 6\mu_1^4 \neq \langle(y - \mu_1)^4\rangle.$$

An advantage in working with cumulants is that for the sum of independent random variables their cumulants sum up. For example, consider two random quantities $A, B$ and the second cumulant of their sum: $\langle(A + B - \langle A\rangle - \langle B\rangle)^2\rangle = \langle(A - \langle A\rangle)^2\rangle + \langle(B - \langle B\rangle)^2\rangle)$, which is true as long as $\langle AB\rangle = \langle A\rangle\langle B\rangle$, i.e., $A, B$ are independent. Therefore, the cumulant generating functions $G$ is a sum. Indeed, it is the log of the generating function of the moments $\langle(A + B)^n\rangle$, which is a product: $\langle\exp z(A + B)\rangle = \langle\exp zA\rangle \langle\exp zB\rangle$.

**A.4:** Growth of entanglement entropy.

Consider an Ising spin chain with the transverse magnetic field in the $x$-direction. The Hamiltonian $\mathcal{H} = \sum_j \sigma_i^z\sigma_{i+1}^z + h\sigma_i^x$. At $t = 0$, the chain is in a pure unentangled state, $\rho(0) = \rho_1 \otimes \rho_2 \otimes \dots$, and all components, $\sigma^x, \sigma^y, \sigma^z$ are nonzero. Find in which order in $t$ entanglement between neighboring sites appears. Use the commutation relation $[\sigma_i^x, \sigma_j^z] = \hbar\delta_{ij}\sigma_i^y$.

**Solution**

$$\sigma_{i+1}^z(t) - \sigma_{i+1}^z(0) = t[H, \sigma_{i+1}^z(0)] + \frac{t^2}{2}[H, [H, \sigma_{i+1}^z(0)]]$$

$$= t\hbar h\sigma_{i+1}^y(0) + \frac{(t\hbar)^2}{2}h\sigma_{i+1}^x(0)\sigma_i^z(0).$$

The entanglement between neighboring sites appears at $t^2$-order.

# INDEX